DEALING WITH HUNGER

DEALING WITH HUNGER

Lord
Walston

THE BODLEY HEAD
LONDON SYDNEY
TORONTO

ISBN 0 370 10464 1
Printed in Great Britain for
The Bodley Head Ltd
9 Bow Street, London WC2E 7AL
by Unwin Brothers Limited, Woking
Set in Monotype Plantin
First published in 1976

CONTENTS

FOREWORD

This book is not intended to be a detailed study of world hunger; nor does it claim to discuss all the ways of overcoming it. For instance it does not mention the possibilities of producing proteins synthetically. It says nothing of the sea or of pisciculture as a source of food. All of these will undoubtedly be of value in adding to food supplies in the future. By the end of the century substantial advances will have been made in the production of synthetic edible proteins. It may well be that the cost of these processes will enable them to compete with proteins from natural sources. If this is so they could serve as valuable additives to the diets of those who today suffer from lack of proteins: in particular this could apply to the diets of children in the developing world. But for those who are accustomed to natural, and especially to animal, proteins it is unlikely that they will prove an acceptable substitute: nor will they be an acceptable substitute to those who wish to improve the gastronomic, as opposed to the nutritional, quality of their diet.

As regards the sea, the supply of fish is limited, and with improved methods of fishing the danger is of over-fishing rather than catching more. The cultivation of vegetable protein in the sea has possibilities, but is still in its infancy. The production of fresh-water fish, on the other hand, can be greatly increased by the adoption of suitable fertiliser treatment and modern breeding methods. In tropical climates, where the natural temperature of lakes and ponds encourages the rapid growth of fishes, the opportunities are great. But even if the production from this source were increased tenfold the total amount available would still be very small in comparison with world needs.

Whatever progress is made in these directions the land will remain by far the most important source of food in the foreseeable future. It is only with more intensive cultivation of the soil that we can hope to conquer hunger. This book, therefore, confines itself to agriculture.

The ensuing chapters set out to place the present world food situation in some historical context, and to show that the probabilities are that by the end of the century hunger and malnutrition will be yet more widespread than they are today. This can be prevented only if there is a fundamental change in the attitude of the rich countries of the world to the allocation of resources. It is then suggested how farmers throughout the world could grow more food, assuming these greater resources were made available. The book has been written in the intervals of a relatively busy life. This may account for some of its shortcomings. But it could never have been completed without the help of many people, to whom I am grateful. Among them I would thank the Ambassadors of Brazil and Cuba, the Chargé d'Affaires of Iran, and the Agricultural Attaché at the Embassy of Israel, who have given valuable information about their countries. The Ministry of Agriculture, Fisheries and Food, the National Farmers' Union, and the London office of the European Economic Commission have all helped with statistical information. Mr Harry Atkinson, OBE, Chairman of the St Lucia Banana Marketing Association and former President of the Windward Islands Banana Growers' Association, has provided much helpful material. To all of these organisations, and to the individuals concerned I am most grateful. I also owe much to my wife and various friends for textual criticisms as well as encouragement: and above all to Vanessa Watkin for her constant typing and retyping, in addition to her help in collecting and checking facts and figures.

I

Introduction

We are told that in the Garden of Eden enough food grew of its own accord without the need to work for it. Be that as it may, ever since then man has known hunger, or feared hunger. In pre-historic days drought or flood could destroy the berries and roots on which he fed, or drive away the animals that he hunted. When he began to cultivate crops and keep animals for his own food there were the same risks. However hard he worked, however skilled and careful his husbandry, the risk was always there. Rarely was the period, in the tropical zones before the rains came, and in those farther north or south, when winter gave way to spring, free from shortage of food.

In those days of close family and tribal ties, these shortages were shared equally by all. Today, too, in small and isolated communities, such sharing takes place. But in the sophisticated and supposedly more advanced societies that are found in most of the world today, this is not the case. We have grown accustomed to the phenomenon of people dying from sickness caused basically by malnutrition in one part of a city while only a few blocks away others are dying from causes stemming from over-eating. Few people in the affluent West, as they enjoy their protein-rich 4,000-calories-a-day diet, give a thought to the tens of millions in other parts of the world subsisting on 1,200 calories, with an almost complete lack of protein.

The history of the change from community responsibility for food and welfare, to indifference for all except oneself and one's immediate family, would make an interesting sociological study. So too would the difference between the blatant exploitation of food shortage, and of famine, in order to acquire yet greater riches for individuals by hoarding food in advance of shortage, and the somewhat more respectable large-scale speculation in commodities; or the even more respectable activities of those who buy and sell on the Futures Market.

But it is outside the scope of this book to study these matters. We shall confine ourselves to the problems of food production and distribution, starting with the period immediately before the Second World War and looking briefly at the food situation as it was at that time in Britain. We shall see first, in broad outline, a country that for many generations had given little thought to food production, and had looked on the world as a whole as a producer of limitless quantities of cheap food waiting to be bought. We shall then see how it adapted itself to a siege economy and came to realise, at least temporarily, that food was not only a pleasure but also a necessity of life and that its production required skill, hard work, and capital investment.

From there we shall move on to the post-war period of food shortages, of the realisation of starvation in closely neighbouring countries, among people who, although at times enemies with whom we were at war, were nevertheless people of the same history, background, and culture as ourselves. From this came the gradual awareness of the fact that in vast areas of the world there were millions who were perpetually the victims of the scarcities that were threatening Europe as a result of a destructive war.

Furthermore, these threats were likely to increase with the rapidly rising number of mouths to be fed throughout the entire world, due, partially at least, to advances in medical science which reduced infant mortality and increased longevity. As a result of this realisation much thought was given, and even some action taken, to increase world food supplies.

But this awakening to the true facts of life was short-lived. Before long the old pattern emerged once more. Apparent food surpluses reappeared, and the talk was once more of restriction of production, and protection for the home producer. Now, some thirty years after the end of the Second World War, even the rich West has been made aware of the existence of food shortages. This threat has been seen as a reality, albeit a minor one, on the shelves of grocers' shops in the cities of Western Europe. The inability of the housewife to buy all the sugar she needs for jam-making, or for the second cup of tea, has not made her and her family aware of what food shortage means for the hungry families of Asia; but it has made her realise that food is not always and inevitably available for her every need.

On the world scene these shortages have brought together many wise and knowledgeable people at the World Food Conference in Rome in 1974, under the auspices of the Food and Agricultural Organisation of the United Nations, and they have revived the talks, the fears, the prognostications and the plans, of the immediate post-war period.

The second part of this book will deal with some of the realities of food production in different parts of the world. It will discuss the theoretical possibilities of greater food production, the practical difficulties that will have to be overcome if theory is to be translated into practice, and the short-term sacrifices that will have to be made in order to achieve long-term security.

The earlier part of the book, as will have been seen, has been written from an essentially European, and to a large extent British, point of view. This has been done not only to concentrate the problem into a more readily comprehensible area, but also because Western Europe, and in particular Britain, have for long been the world's largest market for all primary products. Even today the enlarged European Economic Community devotes one quarter of all its spending outside of the Community on imported foodstuffs. In 1973 the total imports of food, drink and tobacco into the countries of the European Economic Community amounted to $31,670 million, compared with imports into the United States of America of $8,680 million, and into the Union of Soviet Socialist Republics and its partners in Eastern Europe of $6,520. Of these totals the EEC bought $6,950 million's worth from the developing countries, the USA $4,380, and the USSR and its partners $1,630 million. (*Statistics from United Nations Monthly Bulletin of Statistics, Vol. XXIX, No. 7*).

It is clear, therefore, that the policies of the EEC, especially for food, with regard to trade with the Third World will have a profound influence on the standard of living and on the level of nutrition of the people of the developing countries.

II

Before the Second World War

If, in the 1930s, one had asked a typical middle-class Englishman if there were any starvation in Britain he would have replied with an emphatic No. If one had asked him if there were any malnutrition he would probably not have understood what was meant. To him, at that period, one either had enough to eat, and was therefore well-nourished: or else one did not have enough, and was starving. This was a fate that was reserved only for Asiatics and Africans, human beings, of course, but of a very different kind from Europeans and Americans, and above all different from Englishmen.

In those days the science of nutrition was relatively young and the interest taken in such matters confined to very few people. Such statistics as there were would have shown that the average Englishman's intake of food was very near the top of the league table, if not at the very top. They would also have shown that the cost to him of his food, expressed in terms of earning power, was low. This was due to the history of the preceding hundred years.

Until the Napoleonic Wars Britain, in company with every other country of the world, relied for all but a few exotic adjuncts to its food, on its own land and its own farmers. It had, relative to its population, a large area of fertile soil, a kind climate, and an efficient system of land ownership based on the landlord-tenant relationship. Many of its landowners took a deep interest in farming. Some of these travelled widely in other European countries, and they developed contacts with agricultural innovators outside of Britain, and thus introduced on their own estates the best techniques that were available throughout the continent.

As a result Britain became the leading agricultural country of the world, and, in spite of its rapidly growing population (rising from 5½ million in 1688 to 13 million in 1827 and 21 million in 1866 in England and Wales) it was even able to export grain from time to time to other countries in Europe.

With the end of the Napoleonic Wars and the advent of the Industrial Revolution the picture changed. Continental Europe, as a result of the blockade by the British Navy, came to realise that starvation could only be kept at bay by growing its own food. Even sugar, in the 18th century imported mainly for the rich from the Caribbean Islands, came to be produced from the newly developed sugar beet: and, with only a slow growth of industry, there was no shortage of labour for work on the land.

In Britain, opposite factors were at work. The British Navy was in command of the seas, British colonies were growing fast, the British Empire was coming into being, the Industrial Revolution was draining men from the land, and the profits from industry were not only attracting capital away from agriculture but were looking for other and even more advantageous outlets overseas.

Soon the advent of railways opened up the North American continent, and huge areas in South America also. The greater part of the development of South America, and of many other parts of the world too, was made possible by British capital, and the interest on this capital could only be paid with the primary products of the developing countries. Cheap food, produced by low-paid workers, transported along British-owned railways, stored and handled in British-owned silos and cold stores, carried across the oceans in British-owned ships, came in ever-increasing quantities to feed the workers in British factories, and their families. This food not only paid the interest on British overseas investments. It also paid for the imports into these countries of goods from British factories. Furthermore it kept the price of food in British shops low, thereby enabling the employer to pay lower wages and compete with greater ease against his continental rivals in the export market. These growing exports, in their turn, could only be paid for by yet more imports of yet cheaper food.

While these changes in the traditional pattern were taking place the great landowners of the 18th century, who had ruled the country by reason of their acres, their patronage, and their Rotten Boroughs, found their political influence slipping away. In their place were the New Men, the industrialists, coal-owners (some of them great landowners, it is true, but no longer dependent upon the prosperity of farming for their wealth), textile manufacturers, owners of heavy engineering factories. Their growing political influence was directed towards unfettered Free Trade, towards the removal of every obstacle that stood in the way of the export of their steam engines and their cotton piece goods, and the import of cheap food. So British agriculture declined.

More important than the actual fall in production and the passing out of cultivation of formerly productive acres was the decline in the quality of those concerned with agriculture. More will be said of this in the latter part of this book, but it may be asserted that the pre-eminence of British agriculture in the 18th century was due to the fact that farming at that time was both a profitable and respected occupation, and therefore attracted the ablest and most adventurous minds, as well as capital.

Throughout the 19th century, and at least for the first forty years of the 20th century, able and adventurous minds, and capital, went into activities that brought better rewards and greater respect in the community than did agriculture. With a few outstanding exceptions, only those with slight ability or ambition remained on the land. For Britain in isolation this change was serious enough; but it spread throughout the world, so that today the words 'peasant', 'farmer' and 'farm labourer' still carry in most countries a certain amount of contempt; and it is the ambition of most go-ahead young people in developing countries to escape from their rural environment and find themselves jobs in the city.

So it came about that by the 1930s agriculture in Britain was a depressed industry; fields were neglected, buildings were tumbling down, and many villages were becoming deserted. Two-thirds of the food that the country needed was imported and every year, in spite of unemployment in the cities, the drift from the land continued. Farm workers' wages were at the bottom of the scale, and most villages were still without the electricity, main water and sewerage, that were universal in the towns. Education in the countryside was of a lower standard than in the towns, not only because less money was spent on buildings but because most of the best teachers preferred town to country life. And so it was with hospitals and doctors too. No wonder that those who were able to get jobs in towns did so, not only for their own sakes but for the sake of their families too.

It does not follow from this that life in the city was good for the poor. Among the unemployed—and at the height of the depression there were over three million of them—hunger was so real that a march to Parliament was organised—the famous Hunger March from Jarrow in 1932. Voluntary organisations set up soup kitchens to relieve the worst suffering and there were few deaths recorded from actual starvation. But there was widespread evidence of death from cold and sickness, especially among the old. Among children, diseases, which could be traced directly to malnutrition, were common.

This was, in fact, nothing new. The depression and high unemployment brought it into the open, but it had been present before, and it continued after the worst of the depression was over. Rowntree's classic work *Poverty and Progress*, published in 1941, suggested that in the city of York in 1936 31·1 per cent of the working class population were under-nourished; while a study by Boyd-Orr at the same period produced figures to show that 10 per cent of the population had insufficient to eat; a further 20 per cent had enough in quantity but of the wrong quality; and yet another 20 per cent had diets which were not perfect from the nutritional point of view, though this was due to taste or habit rather than poverty.

Thus it can be asserted that even in a relatively prosperous city in the 1930s one-third of the population had insufficient money with which to buy enough food for their family and were suffering, to a greater or lesser extent, from malnutrition. Equally impressive, if less scientific, evidence for this fact can be found in the writings of novelists from Dickens onwards. It cannot therefore be denied that, in spite of the wealth of the British Empire, in spite of the luxurious living of the small minority at the top of the economic scale, in spite of the well-being and comfort of the middle classes, there were still millions of people in prosperous 20th century Britain before the Second World War who did not have enough to eat, and this at a time when wheat was being destroyed in the Middle West of the United States of America, coffee was being burnt in the locomotives of Brazil, the sugar workers of the Caribbean and those who worked in the oil palm plantations of South East Asia were destitute, and fertile acres not only in Britain but all over Europe were being abandoned.

If civilised, and on the whole humane, human beings, like those of the upper and middle classes of Britain, could turn a blind eye to what was happening in their own country, and even in their own towns and villages, while they continued to enjoy the amenities of their own affluent lives, it is indeed a hard task to make the people of the rich countries of the world realise the suffering and hunger that exist in other countries many thousands of miles away. Perhaps it is only the possibility that such hunger may one day in the not too distant future affect them and their children personally that will bring them to an acceptance of the facts as they are today, and, what is more difficult, to an acceptance of those things that have to be done in order to bring about some improvement.

III

Food Policy 1939 - 1945

Towards the end of the 1930s, as the inevitability of war became apparent, thought was given in Britain to the problem of how an island population of over fifty million people, accustomed to importing two-thirds of its food from overseas, could survive the dangers of blockade from submarine and aircraft. Clearly food imports would have to be cut, and what was brought in should be those foods and feeding stuffs which occupied least shipping space. But equally clearly such imports as would continue to come into the country would not be enough to feed the whole population: there would have to be a substantial increase in the amount of food grown at home.

The first line of attack therefore was to increase the production of home agriculture. The second was to ensure by rationing an equitable distribution of all the food available. Both these methods had been used successfully twenty-five years earlier in the First World War. But to them was added a third need—the need to ensure the adequate nutrition of the population, and especially of those, such as children and workers in heavy industry, whose dietary needs were greater than those of the rest of the population.

The policy designed for increasing home food production was a combination of exhortation, education, financial incentive, and penalty. On the purely domestic side, people were encouraged to 'Dig for Victory'—to grow potatoes and other vegetables in gardens which formerly had produced nothing but lawns and flowers. They were also urged to save their household scraps such as potato peelings, stale bread and cabbage leaves, and feed them to chickens or pigs kept in the back yard. At the same time local authorities were encouraged to separate and process all edible waste collected with the normal refuse, and sell it to commercial pig-keepers. By methods such as these a significant amount of food was produced on amenity land, and from what, in more affluent times, were considered waste products; and much that formerly had been thrown away was now consumed.

Such quantities were, however, small compared with the potential of farm land. During the preceding hundred years, with the exception of the short period of the First World War from 1914 to 1918, much good land had gone out of cultivation. Some had been taken for houses and factories, some for roads and railways, and even airfields, some for parks and playing fields. But most had fallen down to grass. This was not the highly productive cultivated grass of a good modern farm, but herbage consisting of natural unimproved varieties, mixed with weeds, producing little of nutritive value. Bushes had taken over in many fields, hedges were overgrown and encroached on the land, drains were blocked and broken and ditches were silted up, so that formerly well-drained land was now water-logged: and rabbits ate a large proportion of such grass as did grow.

Because of the lack of profitability of agriculture in general over many years farms were ill-equipped both with buildings and machinery, and the fertility of much of the soil had been exhausted by over-cropping. Farmers were not only short of money, but were dispirited and lacking in confidence.

In a remarkably short space of time all these handicaps to greater production were overcome. Local committees (War Agricultural Executive Committees) were set up in each county, operating under general guidelines laid down by the Ministry of Agriculture, and consisting of locally experienced and respected representatives of the farming industry—farm workers and land-owners as well as farmers. These committees were charged with the task of increasing production within their own county boundaries, and especially of ensuring that the targets for the main crops such as wheat and potatoes were fulfilled. To help them in achieving this they had powers, within a national plan, to allocate the most important inputs such as machinery, fertilisers, and manpower; and they had the right, rarely used, to remove in extreme cases bad farmers from their land.

In addition to this combination of central and local planning and implementation, the marketing structure of farming was completely altered. The Government, through the newly formed Ministry of Food, became the sole buyer of all the most important crops, at prices which showed a fair profit to farmers without allowing any of the profiteering which is always liable to occur at times of shortages. The cost of inputs was also controlled by the Government, and the banks were encouraged to extend loans to farmers, since more capital was needed with which to finance the increased production. In case these normal channels of finance were not enough, there was provision, not greatly used, for farmers to borrow money from the Government.

The existing Advisory Services were enlarged and a massive programme of education and advice entered upon. This was directed in part to teaching farmers who in the past had done no more than keep a few cattle on poor grassland, how to plough and cultivate arable crops, and how to make the most effective use of fodder and of livestock: but it also concentrated upon the better cultivation and utilisation of grass. Hitherto grass had been looked upon by most farmers as something which grew of its own accord, the farmer having no more to do than put his cattle in to graze it in the spring, and cut some of it for hay in the summer. Now, thanks to the work of Sir George Stapledon and others, grass came to be looked at as a crop requiring just as much attention as any arable crop, with seeds that had to be carefully selected according to soil and climate, and whether the grass was intended primarily for grazing or conservation, and with need for fertilisers and weed control. Also the importance of controlled grazing came to be realised.

As a result of this work of scientists, and of those who spread the new knowledge among farmers, it was found that millions of acres of grassland could be ploughed up and the stored-up fertility used for growing arable crops, while the diminished area of grass, properly treated, could feed as many cattle and sheep as before. In fact between the years 1939 and 1943 the area of permanent grassland in the United Kingdom fell from 18·77 million acres to 11·74 million acres; and, although the number of sheep fell from 26·89 million to 20·38 million, the number of cattle rose from 8·87 million to 9·26 million.

The direct result of all these changes was that food production increased rapidly, so that by 1943 Britain was producing 41 per cent of its total food needs in terms of calories, in place of 31 per cent only four years earlier. The area devoted to wheat was almost doubled, rising from 1·77 million acres in 1938 to 3·46 million acres in 1943. Other grain crops increased from 3·54 million acres to 6·10 million acres, and potatoes from 0·70 million acres to 1·39 million acres.

Total production rose correspondingly. In 1939 British farms produced 1·64 million tons of wheat and in 1943 3·45 million tons. Other grains increased from 2·97 million tons to 5·20 million tons, and potatoes from 5·22 million tons to 9·82 million tons. Put in a different way the land was in 1943 feeding five million more people than it had in 1939, and this with only a small increase in the labour force, which rose from 685,000 to 724,000, many of those being women.

Increased mechanisation helped, though supplies were severely restricted by the war, and so did increased fertiliser use; nitrogen consumption nearly trebled from the 1939 figure of 60,000 tons to 171,000 tons in 1943; the use of P_2O_5 rose from 170,000 tons to 303,000 tons; but, because most of the potash used on British farms came from France and Germany, consumption of K_2O fell from 75,000 tons to 73,000 tons.

The indirect result was no less significant. Almost overnight farmers and farm workers had ceased to be the poor relations of a prosperous industrialised nation, kept alive by the grudging charity of their richer neighbours. They became a vital part of the national economy, receiving a due financial reward for their contribution to the nation's needs, and respected by the rest of the community. The effect of this was far-reaching. Not only were young people, in town as well as country, encouraged to take up farming as a career; either as farm workers or as farmers; engineers found it worthwhile to design and manufacture farm machinery (one of the results of this was that even thirty years after the end of the war Britain remained the largest exporter of agricultural tractors in the world); scientists devoted their skills to agricultural problems, so that the entirely new techniques of weed and disease control in crops by chemical means was evolved; and new varieties of seed were bred which more than doubled the previous highest yields.

The second line of attack, rationing, was the responsibility of the Ministry of Food, which became the sole buyer not only of the main home-produced foods but of all imported food as well. All these were placed on ration, and sold at prices which bore little relationship to the cost of the food but were fixed at a figure which enabled even the poorest to buy their full ration. Those who had formerly eaten well found their standard of feeding lowered: but the poor, and especially the very poor, fed better. What is more, waste was largely eliminated. With only a few ounces of butter, meat and sugar a week nothing was thrown away, even in the most affluent households. To give only one example, before the war the average consumption of meat throughout the country was over 2 lbs a week; but some families could only afford meat once a week, while others ate meat twice a day. In such homes there was significant waste. Under rationing, the average consumption was 1¾ lbs; but every one had the same amount, and nothing was wasted.

This leads to the third line of attack—the nutritional one. Since the First World War the study of nutrition had made great strides, and the Ministry of Food called on the services of one of the leading nutritional experts of the day, Sir Jack Drummond, to advise as to the best means of keeping people healthy even with a restricted war-time diet. Drummond and his colleagues placed the greatest emphasis on milk, and especial efforts were made to increase the production of milk, in spite of a great curtailment of imported feeding stuffs, on which, in the '30s, most of the country's dairy farmers had come to rely. This milk was distributed equally to all the adult population, but an extra allocation was made to children, at school as well as at home. In the schools and in the factories canteens were set up so that factory workers as well as school children received a daily meal in addition to their standard domestic rations. Government-run restaurants were also set up in all towns, where people who were not working in places where there were canteen facilities could supplement their rations cheaply and nutritiously.

Similar facilities were available for those living in villages, or who were sick or too old to go to the restaurants. Special allocations of certain foods such as cheese were made to miners and farm workers and others engaged in heavy work. As a result of the efforts to grow more food at home, and especially to concentrate upon those foods that were nutritionally desirable and that took up the greatest amount of shipping if they had to be imported; and as a result of strict and efficient rationing, the pre-war average consumption of calories of 3,000 per head per day was reduced by less than 5 per cent to 2,860. More important, these were divided equally among all, instead of, as happened before the war, some having far more than the average and some less.

The following tables show that although the composition of the pre-war and wartime diet was somewhat different, from a nutritional point of view, the wartime diet was in no way inferior. Some would assert that it was in fact healthier.

TABLE I

Food supplies in the United Kingdom
in lbs per head per annum (civilian population)

	Pre-war	*1943*
Dairy products (excluding butter) total as milk solids	38·3	50·0
Meat	110·0	86·4
Eggs	28·3	25·6
Oils and Fats	46·9	39·1
Sugar	98·1	66·7
Grain Products	210·1	248·9

TABLE II

Nutrients supplied by above foods:
grams per head per day

	Pre-war	*1943*
Proteins:		
Animal	43·5	39·8
Vegetable	36·8	45·5
Total	80·3	85·3
Fat	130·0	115·3
Carbohydrates	377·5	370·0
Calories	3000·0	2860·0

The effect of such measures was that malnutrition was banished entirely from the country, and the health and growth rate of children in particular was better than it had ever been before.

The lessons to be learnt from this are, firstly, that it is possible to increase food production in a matter of years, even when there is strong competition from others for scarce resources, such as fuel, chemicals, steel, capital and manpower. Secondly, given wise planning of production and a distribution system that pays due regard to nutritional needs, even when food supplies are restricted the health of the population may be rapidly improved. These are lessons which must have especial relevance in those countries where poverty and malnutrition are most frequently found.

IV

After the War

These war-time experiences made people, not only in Continental Europe but also in insular Britain, realise that there could be such a thing as a food problem, and that it could be, at least to some extent, solved by careful planning. Even in the United States of America, untouched by food shortages, far-sighted men were turning their minds to the future supply and distribution of food.

There was, first of all when the war ended, the immediate need to feed the millions in Germany and other parts of Central and Eastern Europe. Under Hitler this whole area had become largely self-sufficient. At the outset of the war Germany was producing more than 80 per cent of its food requirements, and later it was able to draw upon the great food producing areas of France, Poland, Czechoslovakia and Hungary. Coupled with this was an efficient food rationing system. By the end of the war much of this land was devastated, some of it had been abandoned by those who cultivated it; fuel, fertilisers, livestock and machinery were almost non-existent; the rationing system had broken down; and the population had been swollen by millions of refugees from Eastern Europe.

This situation was dealt with partly by massive gifts of food from the United States of America, but also by planned agricultural production in the Western Zones of Germany, the allocation of resources, including steel to farm machinery manufacturers and raw materials to fertiliser factories, as well as by the reimposition of a rigid system of food rationing operated jointly by the Occupying Powers and the German authorities. An indication of the success of these methods was the complete absence of any major epidemic of disease in Europe during the immediate post-war period, which is evidence of the adequate level of nutrition among the population.

Apart from these immedate needs there was concern for the future supplies of food throughout the world. Failure here would mean that there could be no reality for the third of President Roosevelt's famous Four Freedoms—Freedom of Speech, Freedom of Worship, Freedom from Want, and Freedom from Fear. Prominent among those who thought, spoke and wrote about this was John (later Lord) Boyd-Orr, a Scottish agriculturalist, who devoted the latter part of his life to this endeavour. He, and many others too, pointed out that the world population at that time was in the neighbourhood of 2,000 million, and that by the end of the century it would probably rise to 4,000 million. (In fact, already by 1975 it had reached that figure.) Even in the past, with a far lower level of world population, there had scarcely been a period in history when some part of the world had not been struck by famine, and never a time when a substantial proportion of any given community was not suffering from malnutrition. Still worse disasters, and still more widespread malnutrition and the diseases associated with it, could only be avoided by a massive increase in world food production. This could not be achieved unless there were a massive reallocation of resources away from industry and towards agriculture.

This view was opposed by two schools of thought. One maintained that, as in the past, the forces of nature would ensure that an equilibrium was kept between demand and supply; that, when effective demand significantly exceeded supply, war or disease would supervene to reduce the former, or technical advance and an automatic reallocation of resources would increase the latter. The second school developed the second part of this thesis, and claimed that modern methods of farming were already good enough, and foreseeable developments were sufficiently encouraging, to ensure that production would rise fast enough to make certain that world hunger would, at the worst, be no greater than it had been in the past, and, at the best, would gradually be overcome.

The task of this school was made easier by the exaggerated

prophecies of doom enunciated by the more rabid proponents of the need for immediate and substantial action, and especially those who foretold, as a result of soil erosion and the depletion of fertility caused by over-intensive farming methods, an actual diminution in the existing levels of production. In particular they pointed to the Dust Bowl of the Middle West and the destruction of forests in many parts of the world, followed by a significant decrease in rainfall and increase in soil erosion. These factors, they argued, would only lead to a growing loss of cultivated land and the creation of yet more dust bowls and, eventually, deserts. When these dire forebodings failed to materialise within a few years the warnings, not only of the most pessimistic but of the more moderate voices, became discredited.

In the rich countries there was naturally an instinctive preference for the views which encouraged them to continue, after the war, along the path which they had been following with apparent success, and certainly with comfort for many of their people, before 1939. But the very real threat of food shortage was sufficiently close to enable those who were pressing for action to meet with some success. The first, and most significant and long-lasting, of these successes was in the international sphere.

In 1945 the United Nations decided to set up the Food and Agriculture Organisation in Rome, with John Boyd-Orr as its first Director-General. Its task was to stimulate food production throughout the world, and especially in what were then called the under-developed countries, or the Third World. This latter term embraced all those countries which were not firmly bound, politically or economically, either to the capitalist or the communist blocs. Support for the FAO came, naturally, from the under-developed countries themselves. They saw it as a valuable means of lifting their people out of the poverty in which the vast majority of them had always lived; and until this was done they could never attain the full potential of which they were capable.

Support also came from the highly developed countries. In the West there were some who had for long been uneasy at the thought that there still was, in the 20th century, poverty in the midst of plenty. It could not be right, they felt, that millions suffered from lack of food while others ate all they desired, that food was grown but could not be sold, so had to be destroyed, and that large areas of land were left uncultivated. These idealists were now supported by those, equally kind-hearted but of more restricted vision, who until the war had never given a thought to hunger or shortages of food. Now, because of blockade and rationing, and because of the undoubted threat of famine on their very doorsteps in prosperous Europe itself, they realised that hunger did in fact exist, and affected human beings like themselves.

In addition to such people, motivated by idealism, two further categories appeared as allies. The first were business men, who saw that only well-fed and relatively prosperous people could buy the products of their factories, and who therefore wanted to see the inhabitants of the poorer countries of the world with sufficient money to buy the goods manufactured in the West. If every one of these people had more money in their pockets the demand, and hence the profit, would be enormous. Unemployment and bankruptcies would vanish, and the war years would be followed by an indefinite period of expansion and prosperity.

Finally there were those who supported the idea on political grounds. Although the capitalist and communist countries had fought as allies against Nazi-ism, there was still enmity and conflict between those two ideologies. The war against Hitler was followed by the Cold War between East and West, each side trying to gain advantage and allies against the other. The Third World was a vast area where this Cold War could be fought. The West would be at a great disadvantage if it reverted to its prosperous and well-fed life as soon as the war was over, leaving the Third World in the same poverty and hunger that it had experienced before the war.

Just as the seeds of the French and Russian revolutions had been sown by the contrast between the luxury of the aristocracy and the poverty of the peasants, between the ostentation of life in the great house and the poverty in the cottages at its gates, so a revolution on a world-wide scale would become inevitable if the contrast between the West and the Third World remained at existing levels. Something must be done to reduce this difference, to bridge the gap between rich and poor on a world-wide basis. The Food and Agriculture Organisation was the means by which this could be achieved.

So, with a strange combination of motives, this great international agency came into being. Its results, though by now largely taken for granted, have been impressive. It has stimulated research, been instrumental in spreading the results of that research into remote areas of poor countries, and, in co-operation with the World Bank and other bodies, helped to make available some of the investment that is essential if any real progress is to be made in increasing food production. At the same time it has attempted to keep always in the minds of national politicians and the people themselves, the need to produce more food.

The need for greater food production was not confined to action only in the developing countries. Many Western countries, and the Union of Soviet Socialist Republics as well, wanted to grow more food. In Britain, in the immediate post-war period, the need for greater food production even in peace-time was fully accepted, and government policy was directed towards this end.

So far as domestic production was concerned the war-time policies of planned production and guaranteed markets continued, with two changes. In the first place the disciplinary powers of the local committees were gradually relaxed and finally abandoned, on the grounds that once the urgency of wartime had disappeared farmers would not be prepared to put up with the same degree of governmental interference that they had accepted during the war itself. Secondly, as the threat of air and sea blockade was removed, and as shipping space became more available, there was no longer the need to concentrate on growing bulky foods at home.

Now the shortage was foreign exchange, since the massive overseas investments of Britain had largely been liquidated in order to pay for the war; so home agriculture was looked to as a means of saving foreign exchange. This meant that instead of growing more wheat, feed grains and potatoes, the emphasis was now on expensive high-quality foods such as eggs and meat. With these changes the need was still for high domestic production, and it was accepted that this could only be secured by remunerative prices and long-term security. This latter was becoming increasingly necessary as farming became more sophisticated. In the old days before mechanisation and artificial fertilisers farming was largely a question of hard work on the part of the farmer. If it became profitable to grow more he and his family would plough more land, more seed would be held back from the previous harvest, and calves and other livestock retained for subsequent breeding rather than being sent for slaughter. This would eventually produce more meat in the coming years, and would also add to the production of farmyard manure with which to fertilise the additional tillage acreage.

But now farming methods were changing. Tractors and other expensive machinery such as combine harvesters were taking the place of horses, reapers and scythes. Seed was increasingly being bought from those who specialised in the production of pure stock of high-yielding varieties. Artificial fertilisers were supplementing or taking the place of dung. If the farmer were to invest relatively large amounts of money for the purchase of a tractor, usually with borrowed money on which interest had to be paid, he wanted to be certain that the tractor would be used for many days every year. If industrialists were to invest the huge sums needed to build a tractor factory or a fertiliser plant, and if scientists were to spend years of research developing new machines or improved fertilisers, they had to be sure that the results of their investments and of their researches would be needed in the years ahead.

In 1947 therefore the Government brought in its revolutionary Agriculture Act which, for the first time in the history of the country, attempted to give to the farming industry the long-term security and profitability which it had for long been demanding. With hindsight it can be seen that it did not achieve this purpose, since it was based on the assumption that prices would remain stable, and took no account of inflation. In essence it guaranteed to farmers a market for such quantities of all major farm crops, including livestock and livestock products, that it was in the national interest to produce at home: and it further undertook that the price for these products would not be reduced by more than 4 per cent in any one year for any given product, or by 2½ per cent for all the commodities covered by the Act. This had the desired effect, and the farming industry retained that confidence in the future which had been absent in the inter-war years, but which it had acquired during the war, and thus continued to increase its overall production and, above all, make rapid progress towards greater efficiency. The result of this has been that in the thirty years since the end of the war agriculture has had one of the best records of any industry for improved efficiency, both in terms of output per acre and, above all, in output per man.

The Government did not confine its efforts to home food production. It also took steps to increase production overseas, and in particular in countries of the Commonwealth for which it had special responsibilities. Two examples of these efforts show one failure and one success. The failure came to be known as the Groundnut Scheme. At that time all experts were agreed that the greatest shortage in food in the foreseeable future would be in protein, both animal and vegetable. One of the best sources of vegetable protein was the groundnut, or peanut. A survey was made of many areas in the Commonwealth where the groundnut could be grown, and eventually two large areas at Kongwa and Nachingwea, in what was then known as Tanganyika, were decided upon. There were only few local inhabitants in the vicinity, and no adequate means of communication. Therefore it was necessary to build towns and villages, schools and hospitals, airstrips and roads, to set up machinery servicing stations, training schools for mechanics and tractor drivers, and even to construct a railroad to the nearest suitable site for a deep water harbour, which itself was estimated to cost over £1 million, in order to unload the machinery, fertiliser, and other requirements, which had to be brought in, and to take out the crop when eventually harvested.

The original area which it was hoped to clear and on which groundnuts were to be grown was set at 3½ million acres and the total cost was estimated at £25 million. The cost eventually reached a figure of £50 million before the scheme was finally abandoned, by which time only 50,000 acres had been planted, with a yield per acre which was less than half that which had been estimated.

The concept was magnificent. Much-needed food would be grown in an area which hitherto had produced virtually nothing. Education and medicine would be brought to people who hitherto had had none of either, modern agricultural and other skills would be taught, and the example of Kongwa would spread throughout Tanganyika, East Africa, and eventually other parts of Africa also.

Unfortunately the scheme was a failure. Its history has been written elsewhere, and there has been much controversy as to the reasons for its failure. This is not the place to discuss the reasons or to attempt to allocate the blame. Its significance in the context of this book is that it is an indication of the importance that was at that time attached to the need to produce more food: the realisation that great areas of land existed that hitherto had produced nothing, but which, with modern knowledge and the use of methods that had proved successful for military purposes, could make a substantial contribution to the world's supply of food; and the acceptance of the need of massive investment in infrastructure that could only be undertaken by a government or an international agency.

In spite of the failure of the Groundnut Scheme, it could still serve as a valuable example, especially if heed were taken of its mistakes, of what could be done in this field in the years ahead.

The second—successful—example is the Commonwealth Sugar Agreement. As with the domestic agricultural policy this was based on the premise that modern sugar production needed long-term security if it was to be carried out efficiently. It also accepted the contention that there was a moral obligation on the consumer to ensure that those who produced the food he wanted were entitled to a reasonable standard of living, if they were workers, and a reasonable return on capital if they were investors. Thirdly, it was based on the fact that if the inhabitants of Britain were to be assured of adequate quantities of sugar in times of world shortage and bad harvests no less than in times of glut, they would have at certain periods to pay more for their sugar than would be the case if they were to rely solely on unregulated and free markets. This higher price was to be regarded as an insurance premium against shortages.

Under the Commonwealth Sugar Agreements, which ran, with several renewals, from 1953 to 1974, each sugar-producing member of the Commonwealth undertook to sell, and the United Kingdom undertook to buy, a given quantity of sugar annually, at a price varied periodically to cover changes in the cost of production. It was calculated on a basis that would give a fair return to the producer and covered about 55 per cent of the estimated needs of the United Kingdom. About half the balance came from British sugar beet, and the rest was bought on the open market at the ruling price. For the greater part of the time this open market price was well below that of the Commonwealth Sugar Agreement though at times the positions were reversed.

Thus in the 1960s, when the CSA price was approximately £45 per ton, the free market in 1966 fell to £17·87 per ton, and between 1967 and 1971 was around £30 per ton. At one time, however, it exceeded £100 per ton. At the time when it was set up the scheme was a significant step forward in both the theory and the practice of planning world food production in such a way as to hold a fair balance between the conflicting desires of producer and consumer, and so as to give sufficient security to encourage investment and research, without freezing the pattern of production into a too rigid mould.

During the immediate post-war period there were many other attempts to plan production and to stabilise prices of various commodities, including sugar. France had special arrangements for sugar with her colonies and other overseas territories, and the United States of America gave preferential treatment to cane sugar producers in Hawaii, the Philippines and Central America.

More recently, between 1969 and 1973, world sugar production has risen from 69·6 million to 78 million tons, and consumption from 68·4 million tons to 78·7 million tons. Of the total production, about three-quarters are consumed in the countries in which the sugar is grown, and the balance of about 20 million tons is exported. Of this about 10 million tons has been sold under one of three major trading agreements, the Commonwealth Sugar Agreement, the United States Sugar Act, and the Russia-Cuba Trade pact. Thus little more than 10 per cent of total sugar production in the world is sold on the free market. This accounts for the fact that when production exceeds consumption by only 1·2 million tons, as it did in 1969, the free market price is very low; and when consumption exceeds production by only 700,000 tons, as it did in 1974, the price rises to the hitherto unheard of level of over £600 per ton.

During this same period there were also international agreements for wheat, cocoa and coffee. The object of all of these was to prevent excessive fluctuations in price, to protect the consuming countries against inordinately high prices at times of shortage and the producers against a collapse of prices during periods of surplus. The general theory on which all these agreements were based was to fix a range of prices above which producers would not sell and below which consumers would not buy, no matter how much demand exceeded supply or vice versa.

The thinking behind these agreements was good, but in practice they met with difficulties which, at that time at least, proved insuperable. In the first place, if the meetings at which the price range was to be fixed took place at a time when, for instance, the world price of coffee was high, the coffee-producing countries stood out for a price range above that to which the consuming countries would agree: conversely if the talks took place when prices were low the consuming countries demanded a range below that which the producers felt to be reasonable. Even if agreement were eventually reached, there were no means of enforcing it, especially if one major producing or consuming country was not party to the agreement. Thus if the Soviet Union remained outside the Wheat Agreement, or Brazil outside the coffee agreement there was no method of preventing purchases by the one or sales by the other at prices which bore no relationship to those laid down by the agreement. It was carrying idealism too far to think that a country, a large part of whose foreign exchange earnings came from one particular commodity, would forbear to sell an otherwise unsaleable surplus, even at a very low price, if it could find a buyer, but would instead destroy it. Nor was it realistic to imagine that a country that desperately needed a commodity would refuse to offer a price above that laid down in the agreement, especially if it were not a party to the agreement; or that the exporting country would refuse such a tempting offer and sell at a lower price to one of the co-signatories of the agreement.

But the greatest weakness of all these agreements was the absence of an organisation with sufficient funds with which to buy and store the surplus products at the lower end of the agreed price range, and then release them, perhaps several years later, when prices once again rose. At this stage it may be of help to give a short account of some of the attempts that have been made to arrive at an International Wheat Agreement, so that there can be no illusions about the difficulties that lie in the path of such objectives.

In 1927 an International Economic Conference was held in Geneva: at this conference attention was given to the world-wide difficulties of agriculture, but no action was taken till 1930, under the shadow of the Great Depression. After several preliminary meetings an International Wheat Conference was held in London in 1931, attended by eleven wheat exporting countries. The attempts to agree on export quotas failed, largely because of difficulties raised by the United States of America and the Union of Soviet Socialist Republics.

In 1933 a second International Wheat Conference was held, again in London. This time twenty-two countries attended, including importers as well as exporters. The conference resulted in the first International Wheat Agreement, among the provisions of which were the acceptance by Argentina, Australia, Canada, and the United States of America, of export quotas, and an undertaking by them to reduce their exports by 15 per cent. The consuming countries agreed, in certain conditions, to lower their customs tariffs. The agreement failed, largely because, contrary to expectations, wheat prices fell; Argentina, after a big harvest, sold one million tons more than its quota; and importing countries did not lower their tariffs.

By 1938 proposals for a fresh agreement were put forward. From 1935 stocks had been depleted following upon a series of poor harvests in the exporting countries: but good harvests returned in 1938, and the sixteen exporters met again in London to discuss the proposal of the United States of America for an 'ever-normal granary', described as 'a plan to stabilise the amounts of wheat offered on world markets by each nation year after year'. Discussions continued in spite of the war, and in 1942 the International Wheat Council was set up in Washington. The agreement under which the Council came into being aimed at the provision of reserve stocks, control of production, export quotas, and price regulation. The quotas gave to Canada 40 per cent of the export market, to Argentina 25 per cent, to Australia 19 per cent, and to the United States of America, 16 per cent. These countries agreed not to sell below certain minimum prices, and to make available supplies that would not exceed the stipulated maximum. These aims were carried a stage further in 1948, when Australia and Canada and the United States of America also agreed to maintain certain minimum stocks. These same countries undertook to export 13·6 million tons annually at the maximum agreed price if there were a shortage; while thirty-three importing countries undertook to buy the same quantity at the minimum price if there were a surplus. Unfortunately the United States Senate failed to ratify the agreement, so it never came into force, and several countries withdrew.

Between 1949 and 1953 several Agreements on lines similar to those proposed in 1948 came into force, though neither Argentina nor the Union of Soviet Socialist Republics took part. During the four-year period of the 1949 Agreement an average of 14·4 million tons annually were sold under the Agreement's provisions, representing 56 per cent of world trade in wheat. A further agreement ran from 1953–54 to 1955–56, but by then several countries, including the United Kingdom, had withdrawn. During this time the quantity of wheat sold annually under the agreement fell to 7 million tons, representing only 26 per cent of the world trade; and this amount fell still further between 1956–57 and 1958–59 to an average annual figure of 5·4 million tons, or 16 per cent of world trade.

By 1959 it was felt that the scope of the earlier agreements should be extended; and this time the United Kingdom took part in the talks and rejoined the Agreement. In particular it was agreed, not only to attempt to stabilise the world wheat market, but also to promote the expansion of international trade in wheat; to lessen the hardships to producers at times of glut, and to consumers at times of shortage; to help to combat malnutrition throughout the world caused by insufficient consumption of cereals; and the inter-relationship between wheat prices and those of other agricultural products was recognised. Following upon the changes introduced in the new agreement, and also as a result of the return of the United Kingdom, the quantities sold under the agreement rose to 15·5 million tons, representing 36 per cent of world trade.

With various technical modifications these agreements continued until 1968, when a three-year International Grain Agreement was agreed upon, following a conference in Rome in 1967. However, because of the growth of stocks, it was not possible to hold prices above the agreed minimum, and the United Nations Wheat Conference of 1971 was unable to agree upon measures to overcome these difficulties. Whatever the reasons, during the period 1953 to 1972 the price of wheat remained remarkably constant, in spite of substantial variations in stocks. In 1953–54 the closing stocks of the five main exporting countries totalled 51·7 million tons, and in 1971–72 they were 48·8 million tons. In 1965–66 they fell to 33·4 million tons, and reached their peak in 1969–70 of 65·1 million tons.

The picture changed dramatically in 1972–73 when the Union of Soviet Socialist Republics, after a bad harvest, became a large importer of grains. This lead to a fall in the stocks of the five main exporting countries from a peak of 65·1 million tons in 1969–70 to 28·8 million tons in 1972–73, and 23·5 million tons in the following year. At the same time the price of wheat rose from the range of $1.50 to $2.00 per bushel during the '50s and '60s, and a drop to $1.40 in 1969–71, to more than $6.00 per bushel in 1973–74. It is clear from these figures that the modest success achieved by international agreements in the twenty years or so after the war could not be maintained under the stress of major changes in demand, or even minor changes in production.

In this context it is worth remembering that since the end of the war there has been a modest but steady increase in the area of the world devoted to wheat production, and a similar increase in yield, resulting in the total production of wheat rising steadily from the figure of 166·4 million tons in 1949–50 to 368·1 million tons in 1973–74.

No mention has so far been made of general aid that was given during this period by the rich to the developing countries. While the whole idea and execution of aid deserves a lengthy study it does not properly fall within the scope of this book. Aid, however, has in the post-war years played a not inconsiderable part in the fight against hunger. It has been given for a wide variety of purposes. Some of it has been for agriculture itself; some of it for the infrastructure, without which further agricultural growth would not be possible; some of it for industries ancillary to agriculture, such as fertiliser factories; and much of it for purposes concerned with the general development of a country, but with no impact, direct or indirect, on agriculture.

It should, however, be noted that since the end of the Second World War there have been free gifts from the developed to the developing countries that are without precedent in history. In the past rich countries have made investments in poor countries, but solely with the intention of gaining financial benefit from them. Much of the so-called aid given since the war has been on the same basis; much of it has been motivated by enlightened self-interest; but some of it has been given with no idea other than to help the people living in the poorer countries.

Especially in the early days of aid, some of it was foolishly given, and some of it, in the long run, turned out to be actually harmful to the recipient. I remember, for instance, while travelling in Bolivia, seeing an oil-crushing plant put up not many years previously to extract oil from locally grown oil seeds, but, at the time of my visit, abandoned and rusting because the United States of America, in an attempt to help the nutritional situation in the country, had made a gift, under their famous Public Law 480, of a large quantity of margarine, thereby destroying the pioneer industry in the country itself. However, with experience such mistakes were in time avoided, and much benefit has been derived from aid. All the major rich countries have contributed, the largest being the United States of America. Britain's aid has for many years been running at the rate of approximately £250 million a year, and now the European Economic Community is paying special attention to this matter.

At the United Nations Trade and Development Conference in Singapore in 1971 it was agreed that all the rich countries should set themselves a target a figure for aid amounting to one per cent of the Gross National Product: it was a matter of dispute whether this figure should include private investment in the developing countries, or whether it should refer solely to government aid given without any expectation of commercial return. Most of the developing countries held that this target figure should include private investment also. At that time the total official Development Assistance from the seventeen members of the Development Assistance Committee averaged 0·35 per cent of the Gross National Product. Of the larger countries France led with 0·66 per cent of GNP, the United Kingdom was next with 0·41 per cent, followed by Germany with 0·34 per cent, the United States with 0·32 per cent, Japan with 0·23 per cent and Italy with 0·18 per cent. By 1973 the average for all countries had fallen to 0·33 per cent, France still led with 0·60 per cent, followed by the United Kingdom with 0·38 per cent, Germany with 0·37 per cent, the United States and Japan both 0·25 per cent, and Italy 0·14 per cent. While the total amounts given had increased, the contributions as a proportion of

GNP had in most cases declined. At the same time the per capita incomes of the one billion people living in the poorest countries, with per capita incomes of less than $200 per head, had in fact fallen by 0·5 per cent between 1973 and 1974.

It was also agreed at the Singapore Conference that aid should be increasingly given on a multilateral basis, through international agencies, rather than from one government to another, and that a decreasing amount of it should be tied in such a way that it had to spend in the donor country. In fact lending to the developing countries by the World Bank and the International Development Association, which totalled little more than $100 million in 1950, rose to $4,500 in 1974. Of this latter figure nearly $1,500 million was devoted to agriculture and rural development.

Aid of this kind, more properly described as Development Aid, must not be confused with Food Aid. This has at times been used by countries to rid themselves of an embarrassing surplus of a given commodity while at the same time alleviating hunger in poor countries. Increasingly it has become accepted that Food Aid must be given only after careful thought and a full assessment of its implications. For famine relief, and at times of sudden emergency, it is of great value; as part of a concerted world food plan it undoubtedly has a part to play, as will be suggested in a later chapter. But the greatest contribution to increased world food production must, in the long run, come from the developing countries themselves, and wherever food aid is given it must be done in such a way as to avoid such cases as that mentioned earlier concerning vegetable oil in Bolivia, where ephemeral food aid actually hinders or prevents long-term domestic production.

To sum up, it can be said that the years following the end of the Second World War were a period when a greater number of people than ever before became aware of the food shortages which had existed throughout the world for centuries, but to which hitherto few but those actually suffering from them had paid any attention. Much thought was given to different means, technical, economic and political, to overcome these shortages by greater production and improved distribution. Various schemes were evolved in order to implement these objectives, some of which met with considerable success, and many of which failed either through lack of experience, or, more often, through lack of will on the part of the rich countries, who alone had the means that were necessary to ensure success.

Although there were far more failures than successes, in terms of history a major step forward had been taken. There was general agreement that the food supplies of the world were far too important, and far too precarious, to be left to the free play of the markets, and that some form of planning, nationally and internationally, was necessary both for production and for marketing. Those who were active in the fight met with great disappointments, and many of them died believing that they had failed. But the foundations had been laid, even though the superstructure would not be completed for many generations.

V

Feast and Famine

Memories are short, and it was not long before the shortages of the war were forgotten by the people of Western Europe. Agricultural production had been stimulated by the need for more food during and immediately after the war, and technical improvements and the spread of knowledge formerly confined to a few countries, but now, thanks to various forms of technical assistance, available to many others, all helped to increase production.

The Agricultural Advisory Services, which had played such an important part in the drive for more food in Britain during the war, and which, under the name of Extension Services, had for many years been spreading new techniques among the farmers of the United States, were now to be found throughout the whole of Europe, and in the developing countries also. Tractor production in the United States and the United Kingdom had by 1959 risen to a total of 420,000 from a pre-war figure of 185, 000. Before the war the total consumption of artificial fertilisers in these two countries was 8,966 thousand tons; by 1959 it was 26,412 thousand tons. Artificial insemination, coupled with improved methods of feeding and better control of disease had increased milk production in the United Kingdom from 560 gallons per cow, and a total annual production of 1,118·7 million gallons, in 1939 to 735 gallons per cow and a total production of 1,798·4 million gallons by 1959.

World food prices fell and there was now talk of unwanted and unsaleable surpluses. In Britain the prices guaranteed by the 1947 Agriculture Act, instead of being below world prices, thereby enabling people to buy their food at prices lower than if they been allowed to rise to world levels, were now substantially above those of the world market. However, the introduction of the deficiency payment system to farmers enabled consumers to buy food at low world prices while still ensuring to the farmer the price guaranteed by the government. Guaranteed prices and markets were still maintained, though use was made of two significant provisions of the Act. At times prices were reduced by the maximum amount allowed, which, because of slowly rising costs, meant a still greater reduction in real terms. At the same time farmers were reminded of the clause which stipulated that the guarantees would apply only to those quantities which it was in the national interest to produce at home. Because of this proviso the government was entitled to introduce quantitative restrictions on the amounts to which the guarantees should apply; anything in excess of this would have to compete, without protection, on the world market.

In Continental Europe similar pressures were appearing: but here the tradition was different. The lessons of the Napoleonic Wars were still remembered, and it was almost universally accepted that farmers, especially in France and Germany, should have first call on the home market, and that imports should only be allowed if the domestic farmers were unable to meet the needs of the home consumer. In spite of this there were troubles in many areas, with milk being poured down the drains and unsaleable vegetables and fruit blocking the roads.

It was in this atmosphere that the Common Agricultural Policy of the original European Economic Community of the six countries of France, Italy, Germany, Holland, Belgium and Luxembourg was evolved. It was based on the belief that the Community should be largely self-sufficient in all temperate foodstuffs, and that production of semi-tropical crops, such as rice, vegetable oils, and citrus fruits, should be encouraged. A highly complex structure was worked out, with target prices for all major commodities fixed at a level well above those ruling on the world market. If prices within the Community fell by more than a small percentage below these target prices the Community, through an Intervention Board, would step in and buy, storing the product until such time as the price rose once more, or selling them at a loss to countries outside the Community. At the same time imports would be subject to a levy which would bring their price up to the target price.

From the farmer's point of view this policy did not work badly. He knew that he could sell all that he produced, and that if the price fell on the open market there was always the Intervention Board ready to step in. For the consumer, on the other hand, the system was less satisfactory. It meant that he was unable to take advantage of low world prices, and that his food, therefore, cost him more than it did in other countries, and especially Britain, where this system did not apply. It also meant, because of higher food costs, that wages in industry were higher, and that therefore the competitive position of industry in the Community on the world market was weakened. It must, however, be pointed out that, in spite of this apparent handicap, Community exports, and especially those from Germany, continued to thrive at the expense of British exports.

A further, and very serious, handicap of the Community's Agricultural Policy was the cost of intervention buying and storage. Had it only applied to crops which were easily and cheaply stored, and where the world price would soon rise so that the stored stocks could be quickly disposed of without loss, the cost would have been relatively light. But even with wheat, where storage is simple, the cost of maintaining big stocks for more than a few months is great, and large quantities had to be sold at prices which were at times no more than half the figure at which they had been bought from the farmers of the Community. For instance, wheat for which the French farmer had been paid £35 per ton was sold a few months later to the British farmer as cattle feed at a price of £18 per ton. The cost of this fell on the Community budget.

If wheat imposed a strain on the Common Agricultural Policy it was nothing compared to the cost of support for the milk producer. The bulk of the surplus milk was transformed into butter—in itself a costly process—and this butter then had to be stored in expensive cold stores. Eventually the stores became full, and the world price of butter was little more than a third of the cost of the Community butter, even before the cost of storage was taken into account. The Community was forced to sell it on the world market, and the only buyer was the Soviet Union. The loss entailed in this one transaction amounted to £70 million.

In order to avoid a repetition of this situation many incentives were offered to farmers in the Community to move out of milk production—regardless of the enormous unsatisfied demand for milk and milk products throughout the whole of the developing countries—and change over to beef production, for which there was a strong, and, it was forecast, a continuing demand. In the event this forecast was proved to be wrong.

It was at this point that negotiations were taking place for the enlargement of the Community to include the United Kingdom, Denmark and Eire. The United Kingdom was the leading country of those applying to join the Community, and was still wedded to the policy of cheap food. It was also the world's largest importer of food and feeding-stuffs, and had strong ties with, and a feeling of responsibility for, many countries of the Commonwealth.

Especially important to certain Commonwealth countries was the market in Britain for their sugar, and to New Zealand the market for butter. At that time the United Kingdom was importing from New Zealand 189,000 tons of butter a year, under a form of long-term agreement, at a price of £290 per ton, while the Community price of butter was £720. It was finally agreed, after much hard bargaining, that free access to the United Kingdom market for New Zealand butter would be continued for a transitional period, as would free access for New Zealand lamb, one of the largest sources of supply of relatively cheap meat to the United Kingdom and a major export of New Zealand, and that during this period New Zealand would look for alternative outlets. Those outlets proved far easier to find than was anticipated, and, although substantial quantities both of lamb and butter are still imported into the United Kingdom from New Zealand, the price of both has risen substantially and a large demand is developing in Japan and elsewhere in the Far East.

So far as sugar was concerned, the bargaining was even harder. The Community, including the French territories of Martinique and Guadeloupe, was already a small net exporter of sugar, and was anxious, for the sake especially of the French farmers, to increase its production of beet sugar. Britain was the ideal market for this expanded production, and as a full member of the enlarged Community, subscribing to the belief that the home farmer should have the first call on the home market, should only bring in from countries outside the Community such part of its sugar as it could not buy from its partners. But it was unwilling to do so for three reasons. In the first place, as has been said, it had for years provided certain of the poorer members of the Commonwealth with a market for their sugar, which was their main source of overseas earnings, at remunerative prices laid down within the framework of the Commonwealth Sugar Agreement. If these arrangements were abandoned the plight of those Commonwealth countries would be desperate. Secondly, the price of sugar under the Commonwealth Sugar Agreement was at that time approximately £45 per ton (somewhat less for Australia), while the price of sugar from the Community was in excess of £80 per ton. Thirdly, the sugar industry in Britain was a highly complex and integrated one, with Commonwealth sugar imported in the raw form, to be refined in big refineries which could only operate efficiently if there were a large and regular throughput. Some of this sugar was consumed at home, but much of it was re-exported, often to those countries that had produced the sugar, but which did not have sufficiently regular supplies to enable them to refine it economically themselves.

Eventually agreement was reached which allowed Britain to continue to import 1·4 million tons of sugar annually from the Commonwealth, and for the Commonwealth Sugar Agreement to continue until it was due for renegotiation in 1974, and thereafter to be renewed on a Community basis. But here too, as with meat, butter, and wheat, world events supervened, and the picture drastically changed.

It would be very wrong to think that, because Western Europe was experiencing once again surpluses of certain foods, and because prices on the world market were falling, there was enough food throughout the world to feed the entire population. This was no more the case than it had been in the 1930s. Malnutrition was still widespread, and there were still sporadic famines, most of them warranting no more than a short paragraph in Western newspapers and some of them entirely ignored. One such example was in the vast area south of the Sahara desert, where in the seven years from 1967 to 1974 there was almost complete drought which caused crops to fail and dried up the water supplies so that whole communities, with what was left of their livestock, were forced to move long distances in search of water and food.

Even when the rich nations became aware of this situation and mounted an emergency operation of food aid, communications were such that much of the relatively small quantities of food aid that were sent could not be distributed to those areas where the need was greatest. The famine spread to parts of Ethiopia where, it is estimated, hundreds of thousands died; and during the same period there were two serious famines in Bangladesh.

Although food aid was forthcoming in modest quantities for such disasters there was no widespread concern among the rich countries, and indeed no thought that their own food supplies might be affected. But suddenly, in 1973, something happened that caused concern on this score. The price of wheat on the world market doubled, and then trebled, within the space of a few weeks, and there was talk of real shortage. The immediate cause of this was the entry of the Soviet Union into the market on a massive scale, coupled with below average harvests in the Americas. At the same time China increased its purchases, thus putting a further strain on supplies.

Part of this extra demand undoubtedly came from poor harvests both in China and the Soviet Union: some of it, so far as the latter country was concerned, may have come from an expansion of grain-fed livestock, brought about by a greater demand for meat on the part of the whole population. For it is a fact of prime importance in the supply of food in the world that as people's standard of living rises from a low level the first thing they demand is better food. This does not necessarily mean that they consume more calories, but in place of the unconverted carbohydrate-rich foods such as rice, maize, wheat, or potatoes, they feed grain to livestock, thereby converting it into more palatable, and, within limits, more nutritious, protein. But in the process a large part of the calories is lost: in the production of poultry meat it is calculated that one out of every two calories is lost; in the case of pig-meat, four calories of carbohydrate are needed to produce one of protein; and with grain-fed beef eight calories are required to produce one. Therefore wherever the standard of living rises in even a small pocket of the developing world the demand for grain may well double or quadruple.

Whatever the causes of the sudden rise in the price of grains, it continued for the next two years, causing urgent thought to be given to the problem of world food even by those countries which, since the war, had thought themselves immune from such shortages. To the Food and Agricultural Organisation this situation came as no surprise: one of its many tasks had always been to keep a watchful eye on world food trends, and in 1969 it had produced its Indicative World Food Plan for Agricultural Development.

This Plan started as a survey of the world food situation in relation to population growth and general development, and set out to formulate a strategy for combating any worsening of the food deficit which had been revealed in specific quantitative terms for the first time at the First World Food Congress held by the FAO in 1963. It also studied the development of trade in agricultural products. It was intended by these means to help national governments to formulate and implement their own agricultural policies in the light both of probable demand for their agricultural exports and of their own food needs, whether these were to be met by home production or by imports.

Inevitably the forecasts on which such a plan is based are open to question and argument: but it is hard to disagree with the general outlines of the plan and the broad scale of the forecasts.

Thus it starts from the fact that in 1965 1,500 million people were living in the developing countries, of which about 60 per cent lived in Asia, in the densely populated area extending from West Pakistan to South Korea. It is estimated that by 1985 this figure will have increased to 2,500 million. If food supplies were to remain at their present grossly inadequate levels, population growth alone in these countries would demand an increase of 80 per cent above their present production. But if the incomes of these people rose even by a very modest amount, the demand for cereals for direct human consumption would increase by 100 per cent by 1985, and the demand for meat, fish and eggs, by 250 per cent, or 5 per cent and 12½ per cent per annum respectively. At the present time the annual rates of increase in production are 3·1 per cent for cereals and 5·6 per cent for meat, fish and eggs.

When all the various factors are taken into consideration it is estimated that the total demand for food in all developing countries will have increased by 1985 to 140 per cent above the 1962 figure. If production trends in these countries were to continue at present levels there would be a huge shortfall of domestic production, and food would have to be imported from elsewhere in enormous quantities. This would mean increasing demands on the world market, with consequent substantial rises in world prices. The higher prices would benefit the producer of cereals in the exporting countries, which are in the main already highly developed countries, and would hit especially hard the poorest consumers in the developing countries. Even at 1962 prices this would mean an increased expenditure by the developing countries on cereals alone of 7,400 million dollars. If, on the other hand, the extra cereals were to be grown in the developing countries themselves, allowing for increases in yield at the rates of the latter half of the 1960s, an additional 60 million hectares would be needed in Asia to meet the needs of 1985, which is an increase of 50 per cent over the existing area. According to the FAO this could only be achieved at the expense of vegetable crops and of livestock.

An important section of the Indicative World Plan concerns the improvement in the composition of the diet in the developing countries. It stresses the fact that protein deficiencies have serious effects, especially for children; and that they can lead to irreversible effects on mental development, as well as causing more obvious physical disabilities and susceptibility to disease. While an adequate supply of certain vegetable proteins can protect against many of these defects, the main needs are for animal protein in the shape of milk and meat. Here too, as with cereals, the current rate of increase, which is estimated at 1·5 per cent per annum in the developing countries, is insufficient even to meet the increased birth rate; so, without a marked increase in domestic production, and massive imports at a huge cost, estimated at 4,200 million dollars, children in the developing countries will in fact have diets even more deficient in protein in the ten years from 1975 to 1985 than in the preceding decade.

This was the intensely gloomy but honest picture presented by the FAO in 1969. Since then the situation has deteriorated still further, so that, as has already been said, even the rich countries are becoming apprehensive about the future of their own food supplies. The butter mountain, which caused so many headaches to those responsible for the Common Agricultural Policy of the European Economic Community has disappeared: instead there is now a shortage of cheese, and liquid milk supplies are at times barely sufficient to meet the demand. No longer is it a question of making concessions for the Commonwealth sugar producers so as to ensure them access to the United Kingdom market; rather it is a question of begging them to renew their willingness to provide Britain with the 1·4 million tons for which the approval of the Community had been obtained with so much difficulty, and that at a price, not the £45 per ton of the earlier CSA, but at £265, itself less than half the highest price reached in 1974. While in 1975, it looks as if supplies of feed grains such as maize, and of vegetable protein such as soya bean will become somewhat easier than in the preceding two years, prices are confidently expected to remain well above the levels of the late 1960s and early 1970s.

It was in this atmosphere that the World Food Conference

met in Rome in the autumn of 1974 under the auspices of the FAO. Certain hard facts were then placed before the representatives of all the members of the UN gathered there. One of these was that the poorest countries of the world were already facing a serious balance of payments deficit due to the world economic crisis sparked off by the huge rise in oil prices. By the middle of 1975, it appeared, these countries would still be short of 6 million tons of grain, and this grain, to which must be added heavy freight rates, would cost them an additional $1.5 billion, a sum which they could not afford.

Another fact was that world stocks of grain were already at a level below that of the previous year, which at that time was considered dangerously low; and that, even with good harvests, considerable time would be needed to rebuild them to safe levels. The third fact was that, even with substantial help from other countries, it was expected that deaths from starvation in Bangladesh would reach 25,000 in the immediate future; without monthly help of 200,000 tons of food this figure would be greatly exceeded.

A fourth fact was that the developing countries now contain 70 per cent of the world's population but produce only 40 per cent of its food. They do not have the money with which to import more food, nor the equipment with which to produce more from their own soil. Both more food, in the short run, and more equipment, in the long run, can only come from the rich countries, and must come as aid in some form or another. Figures produced by the FAO indicate that the flow of external resources for agricultural development from rich to poor countries would have to rise from the 1974 level of 1.5 billion dollars a year to about 5 billion by 1980 to make such an expansion of food production possible.

Perhaps the most disturbing fact was that even the modest targets of aid agreed by the United Nations General Assembly of 1970, so far from being met, were now farther from being implemented than they were two years previously. Finally delegates were warned that unless the present average rate of increase in food production was accelerated by at least 30 per cent in the developing countries, there would be, by 1985, annual deficits of cereals amounting to 85 million tons. Such deficits are not merely entries in an accountant's ledger. They mean starvation for millions of human beings. How can this terrifying prospect be avoided? How can the world grow more food?

VI

What Must We Do?

The preceding chapters have attempted to give a brief summary of the evolution of thought and of action in the post-war years with regard to food. They have also touched lightly on some of the relevant historical background which has conditioned this thinking. Now it is time to look forward and to see, in the light of past experience and present knowledge, what are the steps that should be taken in the future to bring closer the day when freedom from hunger becomes a reality throughout the whole world.

The policy must, without doubt, be based on ideals: but it must be a practical one, taking full account of the political realities, and not imposing upon politicians the task of persuading those upon whose votes in a democracy they depend to submit to sacrifices in their own standard of living that are manifestly unacceptable.

There are several basic facts to bear in mind. The first is that, even with the existing world population, insufficient food is produced today to ensure an adequate diet for all. Secondly, even if all available food supplies were equally distributed there would not be enough to go round. Thirdly, as people's standard of living rises so they demand higher quality food, and this process of converting simple cereals and other carbohydrates into animal protein is wasteful in terms of calories. Fourthly, the population of the world is rising fast, and may well have doubled itself by the end of the century. At the present time 200,000 more mouths are coming into existence every day, and all these mouths will need, and demand, food.

There are those who think that family planning is the answer to this difficulty. I do not wish to go into the moral arguments for or against birth control. I merely have three points to make. The first is that it must surely be a great reflection on 20th century society if it is unable to evolve a system whereby every normally healthy person who is born into the world today is able to contribute more to the society in which he lives than he takes out of it. This contribution may be in terms of music, literature, painting: it may be in philosophy or science; it may be the production, distribution or exchange of goods; or it may be in agriculture. Since food is the one basic commodity which we all must have, a well-ordered society should see to it that enough of its people occupy themselves with food production to meet its needs, and that abilities that are surplus to these needs can then be used for other forms of socially useful activities. If we say that the only way to solve the world shortage of food is by having fewer mouths to feed we are in fact saying that our society is so badly ordered that we cannot fail to have among us a large number of people who, throughout their life-span, are net consumers rather than net producers, who at the end of their lives leave the world worse off than when they entered it.

The second point is that even if every man and woman between them produced on average no more than two children to take their place when they die the population of the world would continue to rise, due solely to greater longevity. With the advance of medical science and the spread of medical knowledge, life expectancy is steadily increasing: if the average age were to rise from 40 to 60, even with a constant birth rate world population would before long increase by 50 per cent. What is more, the longer people live the greater will be the proportion of people who are too old to work, and the greater will be the effort required from those of working age to look after not only their children but also their parents.

Thirdly, effective family planning requires a certain degree of sophistication in the society in which it operates; and even where this sophistication is present it will take many years before its full impact can be felt. In certain areas of the world, and especially in India where the greatest effort has been made, there are encouraging results. But for every village where advice on birth control can be obtained there are scores where no such advice is available. For every family that benefits from the advice of the clinic there are dozens that ignore it. Eventually family planning may well have some significant effect in slowing down the increase in world population; but between now and the year 2,000, whatever successes it may have, world population inevitably will rise fast. We cannot therefore look to family planning to do more than help in the most marginal way in solving the problem of world hunger.

So far as a more equitable distribution of existing supplies is concerned, this can only be achieved if the minority, who already eat more than is nutritionally necessary for them, eat less in the future. A similar, though restricted, effect would be achieved if they ate less animal protein and more vegetable products. But it is unrealistic to expect that many of such people—and that includes almost all those who will be reading these words—will voluntarily lower the standard of feeding to which they have become accustomed. You, who are right now reading this page, will you forthwith eat less, cut down your intake from its probable present level of 3,500 calories daily, or even more, to the 2,000 or so which is enough for your well-being if, as is probably the case, you are not engaged in heavy manual work? Will you forgo a large part of the meat and the eggs that you daily consume, perhaps to the detriment of your health, in order to make available more food for people living many thousands of miles away whom you have never seen—or even, just round the corner, in the poorer section of your town, whom you see daily?

And if you are not prepared to do this, you who are sufficiently interested in the problem to spend time reading about it, how can you expect others, who give the matter no thought at all, to do so voluntarily? And how can you expect politicians, who are dependent upon the votes and the good will of those same people, deliberately to take action which will make it impossible for them to continue to eat the food to which they are accustomed?

There may well be a small decline in the total amount of food eaten by the rich and the relatively rich in the years ahead, and also a decline in the amount of meat, butter and eggs that they consume, but this will not come about by legislation or by voluntary effort. It will come about either because of advice given by doctors concerning the harmful effects of over-eating and of too much protein: or, more probably, by the rising cost of such foods, which will force a restriction on the amount previously spent on food. But even if this decrease in consumption does in fact take place in a small section of the population its effect on total world food supplies can at the very best be no more than marginal.

Without a decrease in consumption in the developed world there is a place, albeit a minor one, for food aid. It must, however, be recognised that such food aid as has been given by the rich to the poor countries has hitherto come from surpluses and has involved no shortages among those who give. Even when embarrassingly large surpluses do occur they have not always found their way to the hungry mouths of the Third World. For instance the famous butter mountain of the EEC was not given to the poor countries, but sold at a low price to the Soviet Union. The million tons of dried milk that the EEC has in store at the time of writing is not given as food aid, because the cost of the milk itself and of transport would be too great. Instead it remains in store, even though the cost of storage will before long exceed the cost of transport to those countries that so urgently need it. In fact the proposals both for 1975 and 1976 envisage a total of 55,000 tons of skimmed milk powder for food aid,* with a total cost for all food aid from the EEC amounting to slightly more than $200 million for each year. This consists mainly of dairy products and cereals, with a small amount of sugar.

Cereals comprise the major part of world food aid, and by far the largest contributor is the USA. In 1970–71 they gave 8·3 million tonnes out of a total of 12·4 million tonnes, Canada coming next with 1·6 million tonnes, and the EEC third with 1

* Since increased to 200,000 tonnes.

million tonnes. This was at a time of high world stocks and low prices. It is significant that when stocks fell and prices rose in 1973–74 the total dropped to 5·4 million tonnes, the contribution from the USA falling to 2·9 million tonnes, and from Canada to 0·5 million tonnes, while the EEC, now enlarged by three new members, including the UK, gave 0·25 million tonnes more than in 1970–71.

From this it is clear that food aid, though without doubt of some value, is more a means whereby the rich countries rid themselves of surpluses than a means of permanently alleviating malnutrition. It may be that in the future the developed world will set out to grow certain commodities, where it has natural advantages, specifically for the Third World; and that these commodities will be given free, or sold at highly subsidised prices. But even were this done the contribution to world food supplies would be slight.

It follows therefore that the only means of conquering world hunger is to grow more food, and that by far the greater part of this extra production must come from the developing countries. In order to grow more food various factors are needed. Clearly there must be soil, sunlight, and water. There must also be the three 'M's'—men, money, and machinery, including modern techniques: after these comes the fourth 'M'—marketing. All of these will be discussed in some detail in the ensuing chapters: and following upon these will be brief accounts of the methods adopted to increase food production by different countries with different systems of government, during recent years; and an assessment of the results that each of them has achieved. Finally there will be proposals for future policy.

VII

Machinery and Modern Techniques

Availability of land is the first prerequisite of farming. Of the total land surface of the world, a large part is in the grip of perpetual frost, some is bare rock, some is too steep for cultivation. Much of it is desert, which, with adequate water, could produce food; much of it is swampland, or liable to frequent flood, thus rendering it unsuitable for cultivation without expensive drainage. Some of it is rich, well-drained soil, with sufficient water and sunlight to enable crops to be grown on it, but at the present time too remote from centres of population for anyone to wish to live there and grow food. Some of it provides grazing for livestock from its natural herbage; and much of it is already cultivated, but in such a fashion, either because of ignorance or lack of physical resources, that it does not produce more than a very small part of the total of which it is capable.

We shall concern ourselves mainly with those categories of land which are already under cultivation, and those where only relatively modest expenditure on irrigation, drainage and communications is needed to make cultivation possible. It is here that the three 'M's' have a great part to play. Of the three Man is the more important factor; but the thesis will develop more clearly if we start with Machinery and Modern Techniques.

There is a tendency to exaggerate the importance of the part that mechanisation can play in agriculture. In general, mechanisation enables men to carry out the same cultivations on the same land that they have previously done by hand or with the help of animals, thereby saving their muscles and their sweat. In some cases the extra power that is available makes it possible to cultivate soil which is too hard or heavy to be tackled by unaided manpower.

Sometimes it makes it possible for a given amount of work to be done in a shorter space of time than would be the case if no machine were available: in such cases the yield of the crop is increased.

Mechanisation enables one man to cultivate a larger area single-handed than he could manage on his own; but since in most under-productive areas there is already a surplus of labour, a machine that does no more than save labour is not necessarily of benefit. If, however, it is a question of increasing production rapidly without drawing on extra labour, or if increased agricultural production is being attempted at the same time as increased industrial production also, the purely labour-saving aspects of mechanisation are of great value.

In every case machinery makes it possible for a man to do a given job with less effort and removes much of the drudgery. Over the years this has a profound influence upon the relative attraction of farm to industrial or office work, and thereby affects the type of person engaged in agriculture. This theme will be developed in more detail in the next chapter.

To set against these undoubted, but limited, advantages, there are disadvantages. Machinery costs money for the farmer, and he rarely has enough for all his needs. It costs money, too, for the country itself. Unless it be large, it will not have a big enough demand to warrant the construction of a factory, so will have to pay for the machines with scarce foreign exchange. Even if it is large enough, it will have to import much, if not all, of the raw materials, pay royalties to the original designers of the machine, and divert scarce internal resources from other industrial production. The machines, if they be tractors or are powered by internal combustion engines, need fuel, and in virtually every country this has to be imported, thus constituting yet another drain upon foreign exchange.

The machines also need a good supply of spare parts, which necessitates a further locking up of capital: and they need skilled mechanics for their proper maintenance. This is even more important than in their country of origin, for they often have to work in conditions far harder than those envisaged by the original designers.

A further drawback of mechanisation, unless it be of a very simple kind, is that it tends to freeze production into a rigid pattern. To give an over-simplified example, the farmer who cultivates solely with fork, spade and scythe can grow cereals, potatoes, sugar beet, cabbages, hay, with no change of implement. When he invests in a combine harvester he is committed to cereals; if he buys a baler he must make hay, or silage if he buys a forage harvester; if he has a potato harvester he cannot change from potatoes to sugar beet. His freedom of manoeuvre is restricted, even though his efficiency has increased. He is less able to adjust to changing circumstances and increasingly requires stability of markets.

This does not mean that mechanisation does not have an important part to play in an expanded world food production: rather it is to point out that it is far from being a panacea, and that, while still being an essential part of the programme, it brings in its wake problems as well as solutions.

More valuable than mechanisation is the spread of new techniques, the utilisation of new varieties of seeds and planting material, of fertilisers and the chemical control of weeds and diseases. In the 1960s high hopes were centred on the 'Green Revolution' and the introduction of new high-yielding varieties of rice and maize which had been bred in the research centres of the developed world, in particular in the United States of America and Mexico, and which, under conditions of good husbandry, gave yields twice or three times as heavy as did the earlier, unimproved varieties.

Also there was the expectation that the introduction of artificial insemination into the developing countries, together with high-grade foundation female stock, would enormously improve meat and milk production. There can be no denying that such advances substantially increased yields; but the results fell far short of expectations.

For this there were many reasons: a modern high-compression

motor car with fuel injection and automatic transmission is technically a great improvement on the old original Model T Ford, giving greater comfort and higher performance, but it needs high-quality fuel and skilled maintenance, and cannot be kept going by the local mechanic helped with such old pieces of second-hand equipment as he can rescue from the scrap heap. To keep it at its full efficiency he must have access to a whole range of complicated devices for testing and adjusting, as well as spare electrical and other equipment which can only come from a very well-stocked store or a factory. So it is with highly-bred seeds and livestock. Crops, for the full development of their potential, need fertilisers. They must be protected from weed competition by herbicides, and from disease by pesticides. Highly-bred animals must have, if they are to give of their best, well-balanced feeding stuffs. They must be protected from unfavourable conditions by good buildings. If they are not to fall prey to the diseases from which the less highly-bred indigenous animals have developed immunity they must have good veterinary attention, and skilled general management. Without all these things yields of crops and animals will fall far short of what they are capable in their country of origin.

This is not in any way to belittle the enormous importance of high-yielding varieties and improved livestock, especially when used in conjunction with new techniques. Without them progress would be painfully slow, depending solely on bringing into cultivation land which had hitherto been uncultivated. The difficulties have been underlined in order to warn against undue optimism, and to point out the disappointments that are inevitable if it is thought that the introduction of machines and high-yielding seeds is all that is needed in order to achieve rapidly a great increase in output. They are essential prerequisites, but they must be combined with an understanding of all the other things that are needed to obtain full benefit from them. This depends more on the farmer himself than on the scientist and the engineer; but only with such understanding will they be able to make anything approaching the maximum contribution of which they are capable.

The manner in which this combination of modern methods and practical experience can work may be shown by the example of one small Caribbean island. The present banana industry of the Windward Islands was begun in 1954, as a result of a particular form of marketing agreement. The importance of marketing, as has already been said, will be dealt with later. At this stage we are concerned only with the technical aspects of production. In the years immediately preceding 1954 no bananas for export had been produced: within 15 years, by 1969, one island of less than 100,000 inhabitants, and of a total area of about 300 square miles, much of which was uncultivable because of its mountainous nature or unsuitable for bananas because of insufficient rainfall, was producing for export 98,000 tons of bananas annually, and this at a time of slowly rising industrial, and rapidly rising tourist development also.

This was made possible by limited mechanisation; by care in the selection of planting material; by the heavy use of fertilisers and the introduction of chemical methods of weed control; by the recognition and understanding of diseases and therefore the adoption of methods to control them; by a general adoption of methods of cultivation based both on original thought and observation in the field; by research in the laboratories; and by the intelligent use of all modern aids.

How these changes affected not only total production but also productivity and the lives and living standards of the people of the island can be shown by the example of a single plantation. In the second half of the 1950s banana cultivation on this plantation was carried out in the traditional way. The land was prepared and the drains dug by hand (good drainage is an essential part of successful banana cultivation), using no more equipment than fork and cutlass. Planting material was the bulky corm, taken in its entirety from existing plants. For these, big holes, measuring about two feet square and twenty inches deep, were dug by hand in the heavy soil. Weeding was carried out by constant cutting with cutlass.

The only disease control was the dusting of planting material with a chemical to destroy the stem-borer and the spraying of the growing plants with small knapsack sprayers to control leaf-spot. Such small quantities of fertiliser as were applied were carried on the head for long distances to the fields, and the bananas themselves, when harvested, were carried equally long distances to the roadside for selection, packing, and transport by truck to the docks.

Over the years this system was modified so as to make better use of the new aids to cultivation that became available, and in the light of experience and observation. No longer was it necessary to plant an entire corm; instead the active part of the corm, the eye, was cut out and planted, thus saving the carrying of much unnecessary weight. As only a small piece of planting material had to be put into the ground there was no need for a deep hole: instead a small shallow hole, a few inches square and a few inches deep, was sufficient. In this way one man could plant more than four times as much as he could by using the older technique.

Fertiliser was applied in larger quantities, but according to the needs of the plant and the deficiencies in the soil as shown by analyses of the leaves of the growing plant and of the soil itself. Tracks along which trucks and tractors could drive were constructed. (This is an example of the value of mechanisation, for without powerful bulldozers such construction would have been impossible.) As a result there was no longer need to carry the fertiliser long distances on the head, and therefore four times as much fertiliser could be applied by each person as in the past. Similarly the harvested bunch was carried shorter distances to the truck, with a consequent great saving in labour and a lessening of arduous work.

The preparation of the soil was carried out by tractor, as was the drainage, and as a result of a new technique of tractor terracing large areas of land formerly thought unsuitable for cultivation came to be capable of growing heavy crops. Weed control was largely carried out by chemical means instead of cutlass, and leaf-spot was controlled by aerial spraying in place of the former knapsack sprayer. The causes of some old diseases were identified and new diseases coming in the wake of more intensive cultivation were discovered and progress made in their control by sprays.

In addition to all these advances irrigation was introduced. Bananas need for optimum growth two inches of rain every week, and must have at least one inch. At certain times of year rainfall falls far short of these figures. With adequate irrigation, drought was no longer the menace it had been in former years. As a result of all these advances, within a period of five years, from 1969 to 1974, with the area under cultivation expanding slightly, the labour force fell from a peak of 600 people to 120: those who remained had more interesting and less physically tiring jobs, and earned more money, were fed better, and lived in better houses; the others were available for other work in the island, mainly in the road construction sector and the tourist industry, but also in the slowly developing industrialisation.

From a different continent and hemisphere there is an example of similar advance, this time in cattle farming. The following is an extract from a report by a prominent agronomist on a farm in Southern Africa:

> According to the Division of Soil Conservation, Department of Agricultural Technical Services and the local Soil Conservation Committee, the farm was in a parlous state, from the erosion point of view, nine years ago . . . It was grossly over-populated and overstocked and no attention was paid to the deterioration of the land. I was quite prepared to see deterioration due to non-burning but found that with the method of veld management, with high concentrations of stock for short periods to trample the old grass, this has not taken place to any marked extent . . .
>
> In all my long experience of veld management I have not come across a farm before which has not been burned for so long a period . . .
>
> The condition of the stock shows that they have not suffered from this policy of non-burning . . . The old eroded arable lands have been put down to *Eragrestis curvula* for hay and

> I do not suppose there are many farms in the vicinity with such large quantities of conserved food. The lands are not eroding any more with such a good grass cover and the productivity under a sound system of fertilisation and utilisation is steadily rising. The stock being fed on the hay were in good condition . . . Another factor that has also assisted greatly is the extremely clever manner in which the livestock has been managed on a co-operative basis so that it all falls under one system of management and no individual kraal ownership . . .
>
> The training of farm labour has produced excellent results. Inefficient labour on farms is one of the biggest problems with which Agriculture is faced. It was interesting to see how illiterate men could be trained to use mechanical equipment in an intelligent fashion.

It is not suggested that West Indian banana cultivation or Southern African cattle are typical of all kinds of food production, even in the developing farming world: nor is it suggested that a small Caribbean island is typical of those vast areas to which we must look if food production is to be quickly and significantly increased. There are, however, sufficient similarities to allow certain generalisations to be made.

First, there must be adequate scientific knowledge of fertilisers, diseases and plant varieties. Second, there must be appropriate machinery. Third, there must be farmers capable of making use of the new techniques and machines, and modifying traditional methods in order to get the full benefit from them. Fourth, capital must be available to pay for machines, fertilisers and herbicides. Fifth, in many areas there is need for irrigation.

Few parts of the world, apart from certain temperate regions of Europe, have a sufficiently humid climate to guarantee enough moisture for the crops that are grown there, year in, year out. Yet there are vast tracts of land at present uncultivated, or under-cultivated, where enough rain falls during the whole year to ensure adequate supplies if only it were more equally distributed. This rain must be caught and retained for use in times of drought. In some areas this can be done by the construction of relatively cheap small dams, needing no greater equipment than a bulldozer with which to push up earthworks at the mouth of small valleys. In others far more costly and ambitious projects are necessary, with properly engineered concrete barrages. In such cases the cost would be prohibitive if the water were used solely for agricultural irrigation. It must be shared between farming, industrial and urban water supply, and, in the largest projects, electricity generation, on the model of the pioneering Tennessee Valley Authority, begun under the Roosevelt Administration in the United States of America during the days of the Great Depression in the 1930s. This power could well be used in part for the production of nitrogenous fertilisers for use in the surrounding agricultural areas, and it could also form the basis of the industrialisation which so often is a valuable partner with expanding and modernised agriculture.

Whether the scheme be large or small, supplying the needs of a farm of a few hundred acres or an area of thousands of square miles, it must never be forgotten that water is one of the basic requirements of all agricultural production; that many areas are short of water at the time when they need it, but that all except deserts have enough rainfall at some period of the year to provide their twelve months' needs; and that one of the most effective ways of bringing more land into cultivation and increasing the productivity of land that is already cultivated, lies in conserving this rainfall so that it can be used for irrigation in time of drought.

The obverse side of the irrigation coin is drainage. All crops require moisture, but none of them flourish under waterlogged conditions. The soil must be of such a texture as to retain moisture long enough for the nutrients dissolved in it to be absorbed by the plant, but not so long as to kill the tender root fibres by drowning. Some land is naturally free-draining and requires no action by man to prevent it becoming waterlogged. Some is heavy, and the water cannot drain away easily. In times of drought this is an advantage, for reserves of moisture are thereby retained far longer than is the case with lighter soils; but when there is heavy rain this is a disadvantage. It is then that the farmer has to drain his fields.

He can use various methods for this. He can cut open drains, by hand or machine, at frequent intervals, so as to carry away excess moisture; he can, with suitable soil types, draw a mole plough through the soil at a depth of two or three feet, thereby enabling the water to seep down and eventually escape into more widely separated open ditches; or he can lay expensive but effective tile drains, again either by hand or by machine. Whatever method is adopted the cost is relatively low, the benefits to the crop substantial, and the work can be carried out by the farmer himself, with or without machinery.

But there are many areas, and often exceptionally fertile ones, where the level of the land is actually below sea-level, or below the level of the main water-course which eventually discharges into the sea. Holland is the outstanding example of this type of land, but it is also to be found in some of the richest farming areas in England. Before such land can be brought into cultivation extensive and expensive drainage projects have to be undertaken. In its simplest form small drains are dug mechanically, the water then being pumped up to a high enough level to allow it to discharge by gravity into the sea. If the area to be drained is very large, and a long distance from river or sea, there may have to be several such pumping processes before the water eventually reaches the spot whence it can flow by unassisted gravity.

In the past many thousands of square miles too wet for cultivation have been drained and brought into cultivation by such a process. Throughout the world today there exist huge stretches of such swamp land which, with the expenditure of enough money for such drainage schemes, could similarly become highly productive land.

All the methods so far described depend upon the application of existing knowledge; the fullest possible use must be made of what is already known, but by itself it is not enough. New problems are continually arising and there must be a steady feed-back from the farmer to the scientist, so that the latter knows in what direction he must guide his research, just as the engineer must know what are the weaknesses of existing machinery and what new machines are needed in the future. There must be a method of assessing priorities, for no matter what resources are allocated to agricultural research there will never be enough to find all the answers to all the problems: those that will yield the greatest and most lasting results must receive the greatest attention.

When the researcher has completed the first stage of his work there must be places where his provisional findings can be tested under field conditions, and, in particular, tested by the people who will have to use them in practical work. Having passed these tests, there must then be adequate advisory services and demonstration centres to bring to the notice of as wide a range of farmers as possible, in as short a time as possible, the benefits that will accrue to them from the adoption of the new methods or the new machines. When all this has been done, means must be made available to the farmer to adopt these methods or acquire these machines, which in most cases will cost money. Further discussion of this aspect of the problem will take place in a subsequent chapter.

When all these things have been done—and they are all being done at the present time, but on far too small a scale to meet the needs of the future—a big step forward will have been taken in increasing world supplies of food. But they all impose yet greater demands on human resources. The men and women who will make use of them are vital to the success of the programme. It is with the human factor that the next chapter is concerned.

VIII

Men and Women

No matter how efficient the machine, no matter how effective the technique, it is the man (or woman) who works the machine and puts the technique into practice who will determine the degree of success. Even when farming consists of no more than saving seed from last year's harvest, breaking up the soil with a mattock, planting, weeding, and reaping with scythe or sickle, the skilled husbandman will always get a higher yield than the unskilled. The more complicated the machine, the more sophisticated the technique, the greater will be the skill that is needed from the operator.

Yet in our industrialised society the farmer's job is too often looked upon with contempt: peasant, yokel, country bumpkin, clod-hopper may be terms used with a certain amount of affection, but never with respect. The fortunate ones, in most people's eyes, are those who are lucky enough to be born and brought up in the town: if one does not have this good fortune, then anyone who has ambition and drive will make good the handicap as soon as possible by emigrating to the nearest city.

It is not only in industrialised countries that this attitude is found. Even where the entire well-being of the country depends on agriculture those who produce the wealth are held in low esteem. Parents who are ambitious for their children, school teachers who want to see their favourite pupils get on in the world, the girl or boy who aims for advancement and a better life —all try to escape from the soil and the countryside, and find work in the offices or shops of the town.*

*Another example from the West Indies illustrates this: preparation of bananas for shipment is mainly done by women and girls. When this

The reasons of this are manifold and stem from the complex of motives that activate all human beings. With very few exceptions we all wish, for ourselves and for our immediate family, a life which provides security of food and shelter. We then wish for the esteem of the community in which we live. In some societies this esteem will be given to those who excel in some activity which is held in high regard, and which may or may not have economic value. It may be skill in hunting, or in games, in making pots or metalware, in painting pictures or writing poems. The simpler the society the simpler the skills which are highly regarded. In a complex society, such as is found in developed countries and such as is becoming increasingly common in the developing world too, it has become harder to judge of the individual skills of any one person and to balance the value of one skill against another.

A common denominator, and one that is readily visible, has been found, and that is money. Therefore the natural desire of a man for the esteem of his fellows has in most societies today been translated into the desire for money. This possession of money has the additional advantage of being able to ensure the security of food and shelter which is the prime need of all people, as well as the possession of other material things which in themselves are pleasant to have, or which custom, salesmanship or envy, have taught us to desire.

In addition to the esteem of others, whether this esteem be accorded to skills or to money, there is to be found among many people the need for self-fulfilment, the need to use to the full such qualities as they have been fortunate to be born with, or such aptitudes as they have acquired. The strong man wishes to

---

preparation consisted of no more than wrapping the entire stem in cellophane the job was not a popular one. When the hands of the bananas had to be cut from the stems, washed, selected and put in boxes the shelters in which this work was done were called Boxing Plants and acquired the status of an industry. Girls were then anxious to work in them, so that they no longer had to hide from the boys they met on their visits to town that they worked on a plantation but rather could boast that they were factory workers.

excel in sports, the beautiful woman to look yet more beautiful, the artist to produce a masterpiece, the scientist to solve an intractable problem, the acquisitive man to acquire yet more of what he already has.

Others have the desire to exercise power over their fellow men, to control the destinies of others, sometimes because they feel that in this way they can improve the lot of others, sometimes merely for their own gratification. Some are satisfied if this power is exercised over members of their immediate family, their husband or wife or children; others need the wider field of village or city, while still others feel the urge to exercise this power over whole countries or empires. Running through this interwoven tissue of human motives is the contrasting thread of idleness, of the desire to sit back and do nothing, to be fed and housed and have all the major decisions of life made for one by others. In some this thread is predominant: such people may be pleasant neighbours and companions, but their impact on the world is minimal.

Consider now a rural community, in any part of the world, developed or developing. Go to the village school and spend some time among the children there. The majority of them will not strike you as in any way outstanding. They may be clean and tidy and well-behaved, they may be happy and well-fed, but they will have their fair share of that thread of idleness—or perhaps it would be kinder to call it resignation—which enables them to be content with whatever the world offers them. But there will be a few among those children who stand out, children with ambition, with the urge to excel. They may excel at games, or at work; they may hold themselves apart from their fellows or mix with them, not as equals but as leaders. They may make every effort to please their teachers, or go out of their way to rebel against authority.

Few among these children will be content to remain in the village of their birth if all it can offer them is the life of their fathers and grandfathers, tilling the fields and tending the livestock. Even those who are prepared to do so, if they show any trace of above-average ability, will be subject to pressures from parents or teachers to go on to 'better things' and seek their fortunes in the towns. Their highest ambition may well be to become a lawyer. As a lawyer they will make far more money than would ever be possible if they were to remain on the land. They will sit in a clean office instead of sweating in the heat and the dirt. They will return at night to a well-built house with running water and electricity, instead of to an oil-lit shack or cottage, with no water save what is collected in a bucket from a stream half a mile away. They will have a wide choice of companions, rather than being confined to the cramped society of their immediate neighbours; they will have shops and cinemas to visit. And, if their ambitions move in that direction, they will have the opportunity of meeting those who control the destinies of their country, the politicians, and even of becoming a politician themselves.

If they do not set their sights so high, or if their abilities cannot bring them to pass the necessary examinations, they can become bank clerks, shop assistants, government officials. Whatever they decide, or whatever is decided for them, it is certain that few if any of those the visitor picks out as outstanding among the children of that village school will remain to grow food. Even those thoughtful and generous-minded young men who dislike the hustle and competitiveness of the city, and would prefer, if they took only their own desires into account, the quiet life of the village, may feel that they owe it to the girl they are going to marry and to their yet unborn children to work in a town where there are better doctors and hospitals, better teachers and schools, more mental stimulation and less physical toil.

But if full use is to be made of modern agricultural methods the cream of rural youth must not be allowed to leave for the towns: rather the cream of urban youth must be encouraged to look to the country for its greatest fulfilment, as well as for the greatest material rewards and the good life.

In order to bring this about the first essential is that the income of the farmer must rise in relation to that of the factory or office worker. It is not enough merely to aim at a target of equality of earnings between agriculture and industry: if that target were reached it would do no more than stabilise the position as it is today. What is needed, and urgently, is a reversal of the present trend. This can only be brought about by giving to the farmer, and all those engaged in food production, financial rewards that are higher than those they would get elsewhere. In other words there must be a real transfer of wealth from distribution to production, from industry to agriculture, from the urban to the rural areas.

In war-time Britain it was accepted that the free play of the markets was not the most effective mechanism for ensuring food supplies at a time of emergency, and that modern agriculture needed long-term security to enable it to produce food efficiently. Now, in a world threatened by ever-increasing shortages of food, it must similarly be accepted that long-term security for the primary producer is a prerequisite for greater production. In war-time Britain it was also accepted that the food producer's relative share of the national cake would have to be increased, and, inevitably, there would be less cake left over for others who in the past had received larger slices.

This must now be accepted on a world-wide scale. It will not be popular among the industrialised countries, who will see their economic pre-eminence slipping. It will not be popular among the industrialists of the developing countries, who have gained luxury, privilege and power by reason of their factories, agencies, banks and other businesses, and will now have to yield place to the farmers. It will not be popular with the urban workers in any country, who have come to look upon cheap food as an inalienable right, regardless of the poverty of those, not many years ago their neighbours in their home villages, who produce that food. But unless this shift of real wealth takes place, unless the economic prosperity of those who grow the food rises faster than does that of others, hunger will spread throughout the world, attacking not only the poor in far-distant lands but many of those who now think themselves immune from such a threat.

Once this shift in the distribution of the world's wealth has been accepted, and steps, even if slow and halting, taken to implement it, not only will the economic standard of farmers rise but the esteem in which they are held by the rest of the community will rise also. No longer will it be a source of pride for a parent to boast that his son has become a lawyer or a government official, and a source of shame to admit that he has remained a farmer. A parent will try to get his son a job on a farm or in some industry closely allied to farming: the school teacher will encourage his brightest pupils to try for a degree in agriculture rather than in law or accountancy. More young scientists will devote their abilities to plant-breeding or the diseases of animals, more engineers and mechanics will design and maintain farm machinery, build dams and plan drainage schemes. Not only will existing knowledge be spread more rapidly among farmers, and made better use of because of the quality of the new generation of farmers. The pace of advance in knowledge will accelerate as more and better scientists devote their lives to agricultural research.

Financial reward and public esteem will make this possible. But they will not be enough. To the well-to-do urban Western European or citizen of the United States of America country life has great attractions. They can escape from the turmoil of the city to the quiet of the countryside, listen to the birds, watch the flowers blossom in their gardens, and breathe air unpolluted by the exhaust fumes of the internal combustion engine. At the same time they are free to drive in their car to theatres and museums, to shops and offices, to the best schools and hospitals.

But none of this is open to those living in developing countries, or even the remoter parts of Europe or the United States. For them the country is synonymous with primitive life, the town with the comforts and amenities of civilisation. Even if he were to receive more money, even if he were respected by his neighbours for being a farmer, many a man would turn his back on money and esteem if it entailed risk to the health of his family, poor schooling for his children, and a dull life without mental stimulus.

If he is to ask them to share his life as a farmer he must be able to ensure for them good medical services, good schools, and easy contact with the outside world. So the doctors who work in the countryside, and the teachers who teach in the village schools, as well as the buildings in which they work and the equipment with which they are served must be in no way inferior to those of the town.

But why should the brightest young doctors and teachers, who have the pick of the jobs, bury themselves in the country when they have the offer of a job in town? They do not love the land as such and the esteem and the financial reward they command will be no less if they work in town than if they busy themselves in the jungle or the bush. To encourage them to do so there must either be direction by the government as is already taking place in some countries today, or additional pay or the prospects of more rapid promotion; or a combination of all three. There must also be the provision of school buildings, of clinics, of teaching aids, and equipment for medical diagnosis and treatment in no way inferior to those that would be available if they were to work in towns. Also they must have communication with the outside world. They must not feel cut off from new ideas, from stimulating thought, or from those with whom they were students and who are now working elsewhere.

Fifty years ago, even twenty-five years ago, to overcome this would have proved virtually impossible. Before the advent of the jeep or Land Rover, or of the light aircraft, before radio-telephones, radio and television, isolation in remote rural areas was almost complete. It is not so long ago that even in a small and gentle country like England old people could be found who had rarely travelled farther than the nearest town, some ten miles away along a hard road; and that only once a year. In the vast tracts of South America, Africa, Asia, the journey to the nearest town might well take several days on foot or mule, along tracks impassable after rain, over rivers crossed by no bridge but only by fords, across mountains or through jungles.

But now all is, or can be, changed. Roads traversed by motor in all but the wettest season exist or can be built relatively quickly and cheaply with modern machinery. Landing strips for light aircraft can be easily constructed, and contact can be maintained with otherwise isolated settlements by the radio telephone. Radio and television enable the remotest country schools to have lessons identical with those hitherto confined to the large cities, and adults, if they so wish, can have higher education in their homes by the same means. Entertainment too can come into the home in the same way, just as easily as it does in towns. The means are at hand to banish the isolation of the country, and to provide for those who live in it the major part of the amenities of town life. But although the technical means are there, the financial means are still lacking. Governments are still reluctant to spend relatively large sums on providing these facilities for remote areas, where, it must be admitted, the taxpayers' money is being spent in large quantities on a small number of voters, when those with strong political voices are clamouring for the money to be spent on themselves.

It is asking much of any government that has to listen to the voice of the majority to spend large sums in the ways I have suggested; but if life in the countryside is to be made comparable to life in the cities the money must be spent. If it cannot come from the internal resources of the developing countries it must come from elsewhere. If those who grow food are to be as able, intelligent and ambitious as those who work in industry and commerce, the quality of life in the countryside must be in no way inferior to that of the towns. If this is not achieved there can be little of hope of making full use of technical advances in agriculture.

IX
Money

A lot of money, a truly colossal amount of money, will be needed if hunger and famine are to be banished from the world.

More food will have to be grown on land at present in cultivation, and this will entail more machinery and fertilisers and the other requirements of productive agriculture. More land, hitherto uncultivated because of the high costs of drainage or irrigation, or because of remoteness, will have to grow food. To make this possible an enormous expenditure of capital will be needed. More factories will have to be built to produce fertilisers and machines: more scientists will be needed to progress yet faster with research. Those who produce the food will have to receive more for their labour and their skill, in order to make sure that at least as good brains and business ability go in for agriculture as for any other occupation, and the amenities hitherto available only in the cities will have to be brought to the countryside.

It is impossible to say how much this will cost, and to some extent it is irrelevant. As realists we must accept that, whatever the total may be, it cannot all be made available in the space of a few years. All that can be hoped for now is that as much as is politically possible will be forthcoming, and that it is spent on the right projects. Year by year this sum must be increased, and, conceivably, by the end of the century we shall be approaching the correct figure. It is, however, important to have in mind the order of magnitude of the sums involved, before trying to decide where they should come from.

Let us look first of all at the total amount of money spent on food in the EEC. At present this is £50,000 million a year. If the producers of this food were to receive 10 per cent more for their produce than they do at present, the extra total cost to the 250 million people living in the Community would not be 10 per cent more on their food bill, for well over half the cost of the food that the housewife buys in the shops is taken up by processors, wholesalers and retailers. The price the grower receives for the cocoa in a bar of chocolate amounts to less than 10 per cent of the retail price of the chocolate. The figure for coffee and tea, as sold in the shops of Europe, is similar. Even with a simple product like bread the price of wheat accounts for something of the order of 20 per cent of the price of a loaf. The cost to the consumer of a 10 per cent rise to the producer would in fact be between 2 per cent and 3 per cent. As a proportion of the total cost of living it would be unlikely to exceed one per cent. As a proportion of every household's total income it is serious only for the very poorest.

Over the whole Community it could be met in a variety of ways. For the poorest, and especially for the old, there could be food subsidies. In addition to this there is ample scope for economy in our present affluent and wasteful society. For instance some of the inessential costs of processing and distribution could be dispensed with. Milk might have to be collected from the shop by the housewife, in her own container, instead of being put into glass bottles or paper cartons and delivered daily on the doorstep as is the case in some countries. Bread might be sold unwrapped and cut at the table, instead of being pre-sliced and wrapped.

More care could be taken to avoid waste, of which more will be said in a later chapter. If those savings were not enough, others would have to be made elsewhere. Every year, for instance, in the European Economic Community £8,000 million are spent on alcoholic drinks and £6,000 million on tobacco. If the consumption of these items were cut by 5 per cent, if every smoker of 20 cigarettes a day cut his or her smoking to 19, if the average Frenchman cut his wine drinking, the average Belgian his beer drinking, the average Englishman his whisky and gin drinking, by 5 per cent, £700 million would be saved. Some sacrifice would be needed, some change in the national habits that have evolved over the generations would have to be accepted; but in the rich countries of the world there need be no major dislocation of life if the producers of all the food consumed in them received 10 per cent more for their products, and if smoking and drinking were marginally reduced.

In the developing countries themselves, the situation would be different. So far as the farmers are concerned, provided they received 10 per cent more for all they produced they would happily pay 10 per cent more for the relatively small amount of food that they had to buy; and in most developing countries farmers account for more than half the population. As for the others, this change in the relative cost of living between town and country would go some way towards redressing the present unfavourable balance and encouraging more people to continue as farmers: but in some cases special steps would have to be taken on an international scale to help those sections of the urban population who were especially hard hit. But here it must be remembered that the rise in the price of basic foodstuffs such as rice, cereals and sugar which took place in 1973 and 1974 was very much greater than the rise of 10 per cent which is being talked of here.

In general, therefore, it can be said that a rise in the price of food of at least 10 per cent, both in developed and developing countries, could take place without causing undue hardship to consumers anywhere, though certain minor changes in habits, especially among people in the rich countries, would be necessary. Such a rise would probably be enough, provided other costs did not rise correspondingly, to alter the relative positions of farmers and others so as to encourage bright and ambitious young men to become farmers. The money, apart from the special subsidies for the poor and the old, would come directly from the consumers, and no government or international funds would be needed.

Next comes the provision of machinery. The new type of agriculture which will be needed will require a far greater number of tractors and other machinery both in order to produce efficiently and also to take the back-breaking labour out of farming, labour which has so often made people look for a lighter job in the city. The present capacity for making cars in the world is 25 million annually. Assuming it takes 50 per cent more time to produce a tractor than a car, and that it needs 50 per cent more steel, if car production were reduced to 20 million annually, which is the 1975 figure of demand, and the whole of the labour force, raw materials and factory space were transferred to the manufacture of tractors, the production of tractors would be increased by 2·5 million. It cannot be considered a very great hardship for the user of a motor car to keep his car for a few months longer before buying a new one: but in this way the production of tractors could be more than doubled, with plenty of spare capacity and materials for doubling the output of other agricultural machinery also.

When it comes to fertilisers the situation is different. A modern plant producing nitrogeneous fertilisers costs between £50 and £100 million. It is impossible to calculate how much additional nitrogen would be needed to produce the extra food that is envisaged in this book. Some indication of the scale can be gained from those countries that are now trying to become independent of outside sources for their nitrogen. Cuba, for instance, already has one modern fertiliser plant costing about £60 million, and is planning to put up a second. The population of Cuba is between eight and nine million people, and although it imports much of its food it is also a substantial exporter. For the sake of simplicity of calculation let us say that an investment of £100 million is envisaged for a population of 10 million people, or £10 per person. Let us also assume that there are already in existence plants to provide sufficient fertiliser for 1,000 million people. If the population of the world reaches 4,000 million by the end of the century there will have to be a further investment of £30,000 million within the next 25 years, or something over £1,000 million yearly.

This is a substantial figure, but when one remembers that expenditure on defence in the European Economic Community is of the order of £26,250 million a year, in the United States of

America £45,000 million, and in the Union of Soviet Socialist Republics probably about £55,000 million, it becomes clear that a reduction of defence expenditure among these three groups (and China and Japan are excluded) of less than one per cent would provide the necessary capital.

But nitrogen is not the only element that is needed as a fertiliser. There must also be substantial quantities of phosphates and potash, as well as certain other elements, such as manganese in small quantities. The supplies of these throughout the world are large, but not inexhaustible, while nitrogen production requires large amounts of energy which hitherto has come largely from oil, itself also not inexhaustible. With fertilisers, as with so many other things, we have of recent years become spendthrift. They have been relatively cheap to buy and easy to handle as well as giving speedy and visible results, so they have to a large extent replaced the old-fashioned farmyard manure. As the cost of artificial fertiliser rises in relation to other alternatives, so it will become prudent to process what at present is looked upon as waste, and use it as fertiliser.

Many years ago the great agriculturalist Sir Albert Howard, working in India, introduced what came to be known as the Indore process, by which human and animal excreta mixed with vegetable waste was used as a compost on the land, thereby increasing yields substantially both by providing the elements essential to growth and also by improving the texture of the soil by increasing its humus content. In this way it drained more freely in times of heavy rain, and retained moisture longer in times of drought. Modern agriculture must not be so hypnotised by the remarkable results obtained by the wise use of artificial fertilisers as to ignore the importance of waste products as fertiliser. At a time of low fuel prices and low transport costs, and before the introduction of modern methods of handling farmyard manure, it was undoubtedly advantageous for farmers in developed countries to rely increasingly, but not exclusively, on artificial fertilisers for high yields. In the future, even in these same countries not only farmyard manure and slurry, but the processed sewage and other waste from towns and factories may well become economical sources of fertiliser. In the developing countries the balance will be still more in favour of such organic fertilisers, with the consequent substantial saving in investment.

When one turns to the reclamation of land, large-scale irrigation and drainage projects, and above all the provision of infrastructure in the shape of roads, schools, and hospitals, the difficulties of estimating costs are even greater. Each project would need an individual and detailed survey, and the variation in cost in different countries and in different regions of the same country make it impossible to give even an approximate overall figure per acre. Two general points, however, can be made. The first is that today even the poorest of the developing countries are building roads, hospitals and schools. If a larger number of these new constructions took place in rural areas rather than, as is so often the case, in cities, it would be a significant advance in the right direction. At the same time there should be a rigorous curtailment of prestige projects and concentration on small, simple ones.

There should be no more dual-track motorways leading from the capital to the airport, and then petering out in the bush. There should be no more impressive high-rise blocks equipped with automatic lifts and air-conditioning, both pleasant things to have, but not at the expense of schools in the country for the children of farmers. There should even be a limitation on modern hospitals until the country is well equipped with rural clinics and light aircraft for flying doctors. As is being discovered in so many parts of the developed world Big is not synonymous with Beautiful. In fact the opposite is more often the truth.

The second general point concerns priorities. It has already been suggested that a diversion of less than five per cent of defence expenditure by the developed countries could finance the construction of a great number of plants for the manufacture of fertilisers. Hitherto much of the fertiliser used in agriculture has come from the petro-chemical industry. As oil reserves diminish, and as the price of oil remains high in relation to other commodities, investment should be directed increasingly towards other sources of fertiliser. Hydro-electric schemes, especially if coupled with irrigation, can provide the power needed to fix atmospheric nitrogen in such a way that it can be used as fertiliser. In many parts of the developing world geo-thermal heat from volcanoes could similarly be used. Sewage waste from cities could be processed into valuable fertiliser, given sufficient research and investment.

We must ask ourselves if it would not be better to devote more money to projects such as these, even though it would mean some diminution in the enormous sums at present being spent on space research and supersonic travel.

Over a period of ten years, until putting a man on the moon in July 1970, the United States of America spent $25,000 million on space research, and it is probable that the Union of Soviet Republics has spent a similar amount. France and Britain have spent slightly in excess of £1,000 million in developing the Concorde supersonic aircraft, which will enable people to travel across the Atlantic in three hours instead of six. These are great adventures, full of scientific excitement and stimulation. They are leading to the opening up of new frontiers of knowledge, quite apart from the more immediate and practical advantages which accrue from the basic research which is needed to bring them to a successful conclusion. If money, materials, and human skills were limitless no one could feel anything but admiration for them. But all these are lacking in the world today. If they are devoted to such projects they will not be available for other things. If the simple questions were posed 'Would you rather see a man on the moon or save several million people from starvation?: Would you rather halve the time of travel between London and New York, or improve significantly the standard of feeding of hundreds of thousands of children who today are suffering from malnutrition?' there could be only one answer.

So far, in our thoughtlessness and our obsession with 'progress', the question has not been asked. But it must be asked, and unequivocally answered. By all means let us continue with projects of this kind, but let us be sure that we have our priorities and our proportions right. If the money now being spent, and planned to be spent in the years ahead, on aero-space research were halved, significant progress would still be made in this field in the coming decades—but there would be available many thousands of millions of pounds to be spent annually on the infrastructure of the developing countries as well as on health, education, and scientific research. This contribution to the fight against hunger would be incalculable.

To sum up, to make any significant impact on food production there must be a substantial transfer of wealth and of effort to agriculture from other sectors. The food itself will cost more in real terms. Those who have hitherto enjoyed cheap food, at the expense both of the producer and of efficient production and long-term supplies, will have to pay more for what they eat, and therefore will have less to spend on other things. This contribution will come from the consumers direct rather than through government intervention, though it will have political significance. The investment which is needed for factories in which to manufacture the requirements of modern agriculture can come in part from a transfer in demand from the private buyers of cars to the agricultural sector, and in part from the diversion by governments of a small part of their defence expenditure to agriculture—a modern version of beating swords into plough-shares.

To put it another way, it would be a diversion of a small part of the insurance premium at present being paid for the maintenance of world peace through weapons of so-called defence to food for those who in the years ahead may be driven by hunger to acts of violence. The far greater investment needed to make land more productive, to bring fresh land into cultivation, and above all to improve the quality of life for those who grow food, can come from diverting a proportion of the huge sums at present spent on aerospace research and development to these purposes. These last can only come about through direct governmental action. They will require a change in expenditure pattern, but no sacrifice of any kind from any individual.

The FAO has already pointed out the need for some such

investment fund. It is now up to the rich countries, and in particular to the USA, the USSR, and the EEC, to agree not merely on the appropriate reductions in their respective expenditure in these two fields but to hand over the total amounts saved to the developing countries themselves, or to the appropriate international agency—probably the World Bank operating in close collaboration with the FAO—so that they can be spent in the wisest possible way. Another contributing group could well be the oil producing countries, whose revenues have risen so dramatically in the past years.

The World Bank is particularly well qualified to be closely associated with any scheme for further investment in agriculture in the developing countries. According to a paper entitled *Rural Development*, which it published in February 1975, it 'is now the largest single external source of funds for direct investment in agriculture in developing countries. This has resulted from a deliberate shift in the Bank's policy . . . The share of agriculture has increased from 6 per cent of total Bank lending in fiscal 1948–60 to 16 per cent in fiscal 1971–72 and 24 per cent in fiscal 1973–74. The share of agriculture, furthermore, has increased over a period when total lending has expanded several times.'

The Bank has calculated that the investment needed to achieve a 5 per cent increase by small farms is somewhere between $70,000 million and $100,000 million. A large part of this must go to the poorest countries; and it points out that farming is the principal occupation of 75 per cent to 85 per cent of the rural population in Africa and Asia. This rural population is growing at the rate of two per cent a year, in spite of migration to the towns. As a result of 'inadequacies of nutrition, shelter, health standards and other components of a basic level of living . . . rural areas are notable for high levels of morbidity and mortality, especially infant mortality; physical and mental lethargy and inability to sustain hard work on a regular basis; limited ability to recognise or to respond to problems and challenges; lack of awareness; inactive and poor motivation toward improvement and learning; and, often, hostility toward outside sources of change . . . A link between rural poverty and food intake has been established for a number of countries . . . Malnutrition is the largest single contribution to child mortality in these countries . . . It is estimated that 80 per cent of the rural population is completely out of touch with the official health services . . . Access to education can well provide some chance for the rural young to escape from poverty. There are, however, two important considerations which militate against the rural poor receiving satisfactory education. The first is the relative shortage of facilities and the poor quality of education in the rural areas. The second is the relatively high cost of education to the poor in terms of fees, books and other materials.'

Significant progress has been made towards helping developing countries in their agricultural production by the EEC plan for stabilising the export earnings of basic products exported to the Community by the associated African States and Madagascar, and by the associated states of the Commonwealth. The main object of this plan, embodied in the Lomé Convention, is to help these countries in their general economic development rather than to increase food production as such: but since five of the seven main commodities dealt with under the plan are food crops—cotton, groundnuts and groundnut oil, cocoa, coffee and bananas—the beneficial effect on food production will be considerable.

One of the main factors leading to the formulation of this plan is the realisation that wide fluctuations in export earnings on the part of developing countries has serious effects on their economies. When export earnings rise it is not possible quickly to increase investment expenditure, since such expenditure takes time to plan and to complete: the increased income is therefore largely spent on consumer goods. Conversely, when export earnings fall it is difficult to curtail government spending, since long-term commitments have already been entered into. The money to pay for these must therefore, at such times, come either from increased domestic taxation, which imposes severe burdens and political strains at a time when personal incomes are diminished, or from foreign borrowing, which imposes yet further strains on an already weak balance of payments. Furthermore, most of these countries rely to a very large extent for their foreign exchange earnings on only one commodity.

To overcome these difficulties the Community guarantees to each associated state to maintain its earnings from the commodity in question for a given year at a level no less than it earned from the same commodity, on an average, over the five preceding years. If for any reason, such as a drop in price or adverse weather, the total earnings fall below that level, the Community will make up the difference. Provision is also made for the repayment to the Community of some or all of such compensatory payments should the export earnings exceed the average of the five previous years. There is provision for taking into account some of the effects of inflation. The receiving countries also undertake certain obligations, among them being to ensure a degree of stabilisation in the actual price received by the producer individually and also not to divert to countries outside the Community a significant proportion of their total exports, thereby reducing the amount they earn from the Community and thus qualifying for a compensatory payment while increasing their earnings from other countries.

From figures produced by the Community it is interesting to note that had such a scheme been in force between the years 1966 and 1970, the Associated States together would have received by way of compensatory payments, in the case of groundnuts, sums of $4 million in 1966, $10.5 million in 1967, $2 million in 1968, and $9 million in 1970, while in 1969 they would have repaid $6 million. For bananas, receipts would have been $9.75 million in 1966, $16.5 million in 1967, $17.4 million in 1968, $19.8 in 1969, and $24.8 million in 1970. These figures give a good indication of the variation in the export earnings of many developing countries over a very short space of time.

The effect on the individual farmer and on the farming industry as a whole is no less serious than on the economy of the whole country. Just as a government cannot make plans for large-scale development without an assured income from internal and external revenue, so the farmer cannot borrow money for developing or modernising his own farm unless he has reasonable security of markets and prices for several years ahead. Without such security little or no development takes place.

At the same time as ensuring higher prices and stable markets no effort should be spared by the governments of the developed countries to bring home to their people the true facts of the world food position, and to make them understand that the cheap food they have hitherto enjoyed is cheap only by reason of the low standards of life, and of eating, of many of those who produce that food; and that in future, not only out of altruism but out of self-interest too, in order to ensure adequate supplies of food in the future, their food will cost them more. Only by such means will money flow into food production: without money the food that the world needs will not be grown.

X

Marketing, Credit and Co-operation

The preceding chapters have outlined means by which money, machinery and modern techniques can be made available for food production, and how the quality of life can be improved for rural dwellers. But it is not enough to produce food. Once it has been grown it must be brought to the consumer at as little cost, and with as little waste, as possible.

One of the problems that has bedevilled food supplies over the centuries is the recurrent alternation between gluts and shortages, between lean years and years of plenty. In the old days a good harvest meant more food for immediate consumption. It was too much to expect that people who frequently had too little to eat should hold themselves back when there was plenty, solely in order to have a store which would be available several years later if there were a famine. Furthermore it was probable that by the time the food was needed it would have been eaten by rats, destroyed by insects, or stolen by marauders.

Today there are no technical problems to prevent basic foods such as wheat from being stored safely for many years. The main impediment to such storage of strategic stocks is financial. If a million tons are to be stored for several years, money must be found to buy this wheat from the farmer, and the money thus spent will not be repaid till the wheat is eventually sold several years later. In the meantime it will have to pay interest to the lender. Storage silos will have to be built at considerable cost, thus adding still more to the capital required. Hitherto no country or international body has been willing or able to finance such storage on an adequate scale, though, as a result of the World Food Conference of 1974, there are signs that the necessary finance may be forthcoming.

A relatively small amount of the money referred to in the last chapter should be used to build up buffer stocks of certain commodities at times of seasonal or temporary surpluses. The idea of buffer stocks is an old one, but it has never proved effective mainly because of lack of sufficient funds or because of lack of agreement between exporting and importing countries. Wheat, which for purely technical reasons is among the easiest and cheapest crops to store safely, is an example of this. The exporting countries, and in particular the United States, have been unwilling to devote large sums of money to store their surpluses for the benefit of the importing countries, so as to enable them to buy cheaply at times of shortage. The importing countries, on the other hand, and in particular the UK, have similarly been unwilling to devote large sums of money to such storage, which is of immediate benefit to the farmers of the USA by preventing prices from falling to the levels they would otherwise reach.

At the World Food Conference in Rome in 1974 the idea was again welcomed in principle, but once more there was dispute as to who should pay. A World Food Authority, as suggested in Rome, is undoubtedly the correct body to take on such a task: about this there is little dispute. The dispute is about the source of the funds. The proposals put forward in the preceding chapter provide an answer.

With commodities such as wheat and other cereals, rice, and sugar, the problem of storage is relatively easy, in spite of the large sums of money needed both for the construction of adequate storage facilities and for financing the purchase of the surplus quantities. There are, however, other products which are perishable, and which require some form of treatment or processing before storage. Milk, for instance, must be transformed into butter or cheese, or dried: meat must be canned or frozen. Even when the butter has been made or the meat frozen, refrigerated warehouses are required for storage. The cost of storage for both of these is far higher than for grain, as the EEC has learnt to its cost.

Furthermore, with meat, quality deteriorates if it is frozen or canned, as it has to be if it is to be kept for long periods; and the cost of transformation into canned meat, or, in the case of milk, into butter or dried or condensed milk, is substantial. For such products buffer stocks designed to last long enough to be available at times of shortage, possibly several years distant, are not practicable. However, if the people of the rich countries are to continue to have assured supplies of fresh milk, butter and meat at all times, there will be periods of over-production. If the farmer is to have security of markets, and not be left with unsaleable surpluses, these will have to be bought and processed by some form of central agency. It is essential, with such commodities, to have, as part of this intervention buying, a scheme for the rapid disposal of stocks. These should go either free or at subsidised prices to the neediest sections of the population in those countries where the surpluses arise, or to areas where the nutritional need is greatest. This must be done in such a way that the domestic production of the receiving countries is not disrupted. While these occasional surpluses must never be looked upon as part of the World Food Programme they could well be handed over to the organisation responsible for the programme for the distribution of food aid.

It would be unrealistic to expect such schemes of Buffer Stocks and Intervention Buying to apply immediately to all commodities and throughout the whole world. For many years there will be some countries who will wish to have nothing to do with it, and to take their chance in the world market. There will be many commodities which will present difficulties which will only be overcome after considerable experience with simpler crops. As a start there should be international agreement to set up a World Food Authority.

In the initial stage the Authority would not buy or sell on its own account, but only as agent for countries or groups of countries. One of the most important of these groups would be the European Economic Communities, as the world's largest importer of food and feeding stuffs. The EEC should make long-term contracts with the World Food Authority for a large part of its estimated import needs, at basic prices adjusted automatically to take account of input costs and inflation. The World Food Authority would, in its turn, make contracts with the producing countries for those same amounts. It would in addition make further contracts for much larger quantities of the same commodities on behalf of the Agency responsible for food aid. This food would be made available, at full price where possible but otherwise at subsidised prices, to the food deficiency countries of the developing world.

The second stage of the World Food Authority's activities would be to buy at agreed intervention prices, the surpluses of certain commodities as these arose, and to store them for use when supplies were short and when, in the absence of such buffer stocks, world prices would rise: or, in the case of perishable commodities it would pass them on to the Food Aid Agency. For some products there would be an open-ended commitment to buy all that came on to the market: for others the guarantee would be limited to specific quantities.

It would be the job of the World Food Authority to allocate the amounts covered by guarantee between the different countries which agreed to participate in the scheme. These proposals are not significantly different from those made at various times since the end of the Second World War for many commodities such as wheat, coffee, cocoa and sugar, all of which failed to a greater or lesser extent because of lack of adequate funds. If funds were made available on the scale envisaged in the preceding chapter there would be no problem on this score.

Efficient marketing, at farm as well as national level, will also assume a growing importance. In the days when farmers grew at their own risk and took their produce to the nearest market or merchant where they obtained the best price they could for it, the responsibility was theirs to see to the quality of the product. When a crop is produced mainly for export, or for transport to a more distant centre, and when the farmer is guaranteed a market and a price for whatever he grows, the question of quality takes on a far greater importance. If he is a milk producer, and the milk is collected from his farm, taken to a central dairy, and mixed with the milk of a hundred other dairy farmers, it is essential that all the milk that comes into the central dairy is of a minimum standard both as regards content and also cleanliness.

If he is a rice or cereal farmer and the crop is collected and stored in central silos it is again essential that the grain be of a more of less uniform moisture content and free from impurities and disease. This is even more important when dealing with fruit or vegetables, which can deteriorate rapidly in transit if improperly packed or harvested in bad condition. Few farmers are qualified to control the quality of their products so as to ensure a standard high enough to satisfy the requirements of central marketing, and especially of export. Yet the system suggested earlier of guaranteeing markets and prices is based on just such central marketing. It will therefore be necessary to set up an organisation responsible specifically for advising as to the best means of obtaining a high quality product, of inspecting in order to ensure that the requisite quality is achieved, and with power to reject anything that does not reach this standard. It matters little if this body is under direct government control or if it is organised by a statutory Marketing Board or on a local or co-operative basis. It must be competent, honest, and consist of people with practical experience and able to talk to farmers with authority.

In these ways stability and security would be brought into the world market: but this stability and this security must be passed on to the farmer himself if production is to increase. Unless he knows without doubt that he will be able to sell the crop he produces at a price that will show him a profit he will not make the investments which are needed if full advantage is to be taken of modern methods. Mention has already been made of the need for more mechanisation and fuller use of fertilisers and other aids to high production and productivity. It has also been shown how finance can be provided to ensure that all these aids to efficient farming can be produced in adequate quantities. But there is no point making more tractors and more fertilisers if the farmer is unwilling or unable to buy them.

Farmers throughout the world are rarely organised in the same way that businesses are—they are not conversant with accounts, other than the simple fact as to whether they are more or less in debt at the end of the year than at the beginning. They do not feel at ease when they visit the bank manager, the moneylender, or the merchant, and ask for yet more credit. They do not employ accountants to advise them or have the benefit of a financial director on their board. They are therefore unwilling borrowers, having been brought up in the age-old belief, born of long and bitter experience, that once they incur a debt it is very hard ever to discharge it fully. They have also learnt that even when a crop has eventually been successfully harvested it may not be possible to sell it; and that, in general, while a good harvest means more food for themselves and their families, it usually means a lower price in the market. To persuade such people to borrow what to them is a large sum of money in order to buy fertilisers which will increase their harvest is not easy: to persuade them to borrow yet more money with which to buy a tractor, when they fear that even with a bigger harvest they will not earn enough money to enable them to discharge the debt, is harder still.

This will only be possible if each farmer has confidence that he can sell his crop, not only for one year but for several years ahead, and that he will receive for it a price that will give him a profit high enough to enable him to pay off the debts he has incurred in growing it, and have something left over for himself. Once this has been done he will then invest money and expand production.

How this works in practice can be seen by returning, in somewhat more detail, to the example already given of the banana industry in the West Indies. Several attempts had been made in the inter-war years to start a banana exporting business in the Windward Islands. The soil and climate were suitable, there was ample land and labour, and there were few rival crops. But all attempts had failed because there was no guarantee of security. In 1954 a fresh attempt was made. The four Windward Islands were offered a contract, the two salient points of which were that the buyers would undertake for the next fifteen years to buy all the bananas produced in those islands, regardless of quantity, and to pay for them a price based on that received by the already well-established banana growers of Jamaica. The two provisos were that during that period bananas were to be sold to no one else; and that they would have to conform to certain standards of quality.

In order to implement this contract each island government legislated to set up a marketing board, which had a monopoly of buying and selling all bananas in that island destined for export. Every producer was compelled to sell to the Board, and the Board made the contract with the buyer. It was also responsible for controlling the quality of the fruit and for a certain amount of disease control. As a result, the export of bananas rose in the four islands from zero, in the years prior to 1954, to a total of 180,000 tons by the end of the fifteen year period to 1969. Fertiliser consumption in the largest of the islands jumped from 1,000 tons in 1954 to 10,000 tons in 1969, while tractor imports doubled in the same period. Earnings from exports rose to the £4 million mark, and there was a notable improvement both in housing and in nutrition due to the increased spending power of the banana grower.

From this small example it is clear that, given long-term security and remunerative prices, growers in developing countries will respond rapidly, will increase production, and will borrow money with which to finance this expansion and increase their efficiency.

When the first hurdle, that of the farmer's reluctance to borrow, has been cleared, there remains one problem. From whom is the money to be borrowed? The sources of credit in many developing countries are few and expensive. Traditionally, money comes from three sources, the moneylender, the merchant, and the landowner, and frequently these three sources are found in one and the same person. The moneylender, if he is a separate person, is usually the last resort. More often money is borrowed from one of the other two.

Although the needs of the peasant farmer in a primitive society are few, there are always some things that cannot be met from his own land or labour, and that have to be bought. He must go to the merchant for the oil for his lamp and for his salt; he must buy some meat or fish, cooking utensils, and clothing and shoes, if he can afford them. Modest though his purchases are there must be money to pay for them, and they must be bought before the harvest has been gathered and sold. So he buys on credit from the local shopkeeper, undertaking to repay out of his harvest and often undertaking to sell his harvest through the same shopkeeper, who is also the local merchant. This man will make a profit on what he sells in his shop, and a profit on the farmer's produce which he buys: he will also charge what may well be an exorbitant rate of interest on the loan he advances. The harvest is therefore often all consumed by those charges even before it is gathered; and a fresh debt will be contracted during the following year.

The situation is often the same with the landowner, even though he be not the merchant (and often the merchant has become landowner through taking over the land of farmers who have been unable to discharge their debts). He exacts for the use of his land either a rent in cash, or, more usually, a share of the crop, or a fixed amount of the crop. If the harvest is good and the prices high (a rare combination), the farmer has enough with which to pay his rent, keep enough for his own use, and sell enough to raise money for his needs during the coming year. But when this is not the case he must once again borrow against his harvest. So the landlord takes not only the share to which he is entitled under the share-cropping agreement, but a further share to cover the money that he has advanced during the year.

The moneylender who is neither merchant nor landowner is only resorted to if the other two sources of credit have been exhausted. Consequently, as no security can be offered for the loan, the rates of interest are even higher than those demanded by the other two. He, even less than merchant or landlord, cannot be looked on as the man to go to if a substantial loan is needed for fertiliser or tractor. Here arises the need of some other organisation, with adequate funds at its disposal and prepared to lend at low rates of interest.

Sometimes established banks may be willing to do so, but in many cases they are not eager to set up branches in the country when they already have difficulty in finding suitable staff for their expanding business in the cities. In any case they will naturally demand commercial rates of interest which, while far lower than those exacted by the moneylender, will be substantial. To provide this capital, therefore, there will be needed some form of agricultural credit organisation, designed specifically to meet the needs of farmers, and provided with funds from government or international sources which will enable it to lend on easier terms, both as to interest rates and repayment, than is possible for a purely commercial undertaking. There are many ways in which this can be done, but the three main ones are through a direct government bank, a co-operative credit organisation, or credit facilities made available through the marketing organisation set up to handle the crop in question.

It matters not which particular method is chosen so long as, subject to the lender satisfying himself as to the technical and commercial viability of the project, the farmer can get credit, and get it cheaply and easily; for it must be stressed again that the farmer is instinctively an unwilling borrower, while it is in the general interest that he should borrow. In other words the usual situation is here reversed. The bank manager normally sits in his office waiting for would-be borrowers to come to him, and when they do he must subject them to a searching examination not only as to the prospects for success of the venture for which they wish to borrow money but also as to the security they can offer in order to cover the loan if the venture fails.

In the case of the farmer in a developing country, on the other hand, his unwillingness to borrow and his fear of getting into debt must be overcome; the lender must be an aggressive salesman as well as a banker. In few cases will the borrower have any security to offer other than his skill as a farmer, his willingness to work hard, and the certainty of a market and a fair price when harvest comes. The main security will in fact be his own personality. For these reasons the man who makes the final decision about the loan must above all be a good judge of character: he must also have a sound knowledge of the particular form of farming that is practised in his district, so that he can decide for himself whether soil and climate are suitable for the crop that it is proposed to grow, and whether the particular machine that it is proposed to buy will help in the production. It may well be that in order to do this he will want to call upon the help of someone with more specialised training in the crop in question and in agriculture in general than he could be expected to have. He must also, of course, be well-acquainted with the marketing prospects of the crop, and know the details of the national and international marketing arrangements.

There must therefore be close co-operation between whatever organisation—be it State Trading Board, Marketing Board, Producers' Co-operative—which is set up to organise the marketing of the product, the technical advisory services which are there to advise farmers as to the best practical means of improving and increasing their production, and the providers of credit. In some cases a National Agricultural Bank could undertake all three operations; in others they could be in large measure, as in some countries they already are, performed by private commercial banks in so far as the last two are concerned; in yet others, statutory marketing boards can do all that is necessary; or it may be that agricultural co-operatives, operating on a regional basis and dealing with most of the different crops grown by the farmers of the district rather than with only one crop grown throughout the whole country, will prove most effective.

Whichever method is chosen, the essential thing is that the man on the spot, call him bank manager or what you will, must be prepared to go out and sell credit to a reluctant farmer, that he personally, or helped by technically qualified colleagues, must have knowledge of the practical problems of the farmer, and that he must have very wide discretion in making his decisions and not have to refer all requests for credit to a remote board of directors who spend their time sitting in offices and never get their boots dirty.

The modern farmer, whether in a developed or developing country, cannot be self-sufficient or entirely independent of others. He will not only need help in marketing his crops, and help in borrowing the money he must pay for the machines and other inputs that he will increasingly require. He will also need help in order to learn a whole variety of new techniques. These techniques will be mainly concerned with the means of producing the crop, but they will also deal with marketing it in good condition and of the right quality, and with assessing the profitability or otherwise of different methods and of alternative expenditures. Even a highly educated man with a degree in agriculture and many years of experience cannot expect to do all these things successfully on his own, and at the same time to run his own farm and keep abreast of new ideas. The problem becomes all the greater if he is dealing with more than one crop, or with livestock as well as crops.

For these reasons, in addition to the need for research scientists working in laboratories and on experimental farms, there must be an advisory service of 'dirty boot experts', trained at universities or specialist farm institutes as agriculturalists, but also with practical experience, who spend a small part of their time keeping in touch with new developments in agricultural science, who have contact with specialists in such fields as plant diseases, animal feeding, veterinary science and mechanisation, so that any problems they come across can quickly be referred to the appropriate specialist, but who spend most of their time visiting farmers, walking round the farm with them, seeing the crops and the animals with their own eyes, and giving sound advice.

This body of men must be backed up by one or more commercial farms, possibly run by the Department of Agriculture, perhaps by the marketing organisation or co-operative, or even by private farmers who have been selected because of their ability and interest in new methods, to which the farmers of the district can come, and see the success—or sometimes failure—of what has been tried out. Such advisory officers should also be able to help farmers with the preparation of their accounts, and in particular to advise whether a proposed expenditure is likely to yield a worthwhile return. It might well be to those people that the lender could turn when there was a question of a loan. The task of such advisers would be greatly facilitated by making use of films, which can be shown in local centres, to demonstrate new ideas and methods, as well as by a co-ordinated programme, on a national or regional basis, of educational items on radio and television.

Hitherto farmers have been referred to as individuals. The solution of their problems, the easing of their difficulties, have been discussed as problems and difficulties appertaining to each farmer on his own. This is surely the right approach, for in all farmers there is a strong streak of independence. They are no less good than anyone else at getting on well with their neighbours, they are no less community-minded than are people living or working in towns, but they cherish their independence in a way that is not always found among those who work in factories. For all that, while retaining their independence, they can still find many advantages in co-operation with each other. Marketing is one such sphere.

From the point of view of the buyer, it is clearly of advantage to be able to deal with a group of farmers, each producing a small quantity of a given product, rather than having to make separate bargains or contracts with many small producers. If the group is prepared to undertake control of quality also, so much the better; and so much higher is the price that the buyer is prepared to pay. On the buying side the advantages are no less great. The seller of fertilisers will offer better terms if he can sell an entire truck load of fertiliser—ten tons, delivered to one central point, with one invoice, and the collective guarantee of payment of a group of farmers, rather than having to deliver a dozen or more lots of less than a ton each, collect payment from a dozen or more customers, and run the risk that one of them may delay payment or even default.

Such a farmer's group, or co-operative, could also afford to employ, even on a part-time basis, an accountant who would not only collect payment for all that the co-operative sold, and distribute the amounts in appropriate proportions among its members, but could also keep the accounts of the individual members and help them with their financial problems. If the co-operative were big enough, and especially if its members concentrated on one product, it could even employ a technical specialist to advise on actual production. Whether large or small, credit is always easier to obtain for a co-operative than for individual farmers; for the risk to the lender is then spread among many, all of whom share the responsibility for meeting interest charges and repayment, rather than resting on one man who may be incompetent, have bad luck, become sick or die, or conceivably be dishonest.

When it comes to the co-operative ownership of machinery the benefits are still there, but the difficulties are greater. It is rare, except on the largest farms, for a machine to be used to the limit of its capacity; but the farmer wants to use it at the very moment when, in his opinion, conditions are right, or when he has the time to do the job. He will not be content to wait until one of his neighbours has finished with it. Furthermore, a machine will not last as long, and will cost more for maintenance and repairs, if it is operated by many different people, rather than being in the sole charge of one man who has full responsibility for it.

There are, however, certain types of machine which are too expensive and specialised to be owned economically by any but the largest farms, and where the need is not an urgent one, but the work can be done at any time of the year, or during a prolonged season. Examples of this include a big crawler for terracing hillsides, a specialised machine for digging or cleaning out ditches and irrigation channels, even machines for harvesting maize or making it into silage. None of these can be fully employed on any normal farm, and the period when they are needed is an extensive one. Here there is scope for co-operative ownership. But where such is the case there must be one man, whether a member of the co-operative or an employee, who is responsible for operating and maintaining the machine; and a system of allocation which avoids unfairness and friction.

Co-operation has much to offer, especially when farmers who have been accustomed to simple unmechanised farming methods are brought rapidly into modern agriculture. But there are many difficulties, quite apart from the farmers' natural desire for independence. Success is more likely when the farmers who form the co-operative all know each other well and come from the same background, have the same education, and have approximately the same economic standing. If some are big, rich farmers and some are small and poor, if some have higher education and some are barely literate, if some are newcomers to the district and others have lived there from time immemorial, failure is almost inevitable. This may well be the reason why the co-operative movement, which was started in England in the first half of the 19th century and from there has spread throughout the whole world, has, in Britain, met with such poor success in agriculture, for the composition of the typical English village is widely varied in social, educational, and economic background. In France and in Germany, on the other hand, and in Holland and in Denmark, agricultural co-operatives are an important and valuable part of the rural scene. Where other services are already well established, where there is already a satisfactory marketing system, where honest merchants are easily found, where there are good advisory services, where there are banks and professional accountants, the need for co-operatives is less. In most developing countries none of these services are readily available. There is therefore scope for farmers' co-operatives. It must be a matter of local preference, of political decision, of tradition, and of types of crops that are grown and methods of marketing that are already in practice, which will be the determining factors.

Whether the services are provided by government, by statutory marketing boards or by co-operatives, the farmer of the future must have easy access to credit, to marketing facilities, to technical advice, to accountants, and to machinery. Without them progress cannot fail to be lamentably slow.

XI

Landownership

There are many different forms of landownership, and they all have some effect, good or bad, upon the success or failure of different methods of cultivation. In order to get the best possible results from modern farming techniques it is important to ensure that the right forms of ownership are adopted and that those with manifest drawbacks are discarded.

In the earliest times, and even today in certain simple communities, the ownership of land was of no significance. There was no cultivation and there was land in plenty. People lived on wild berries and roots, and hunted animals and such domesticated animals as they had grazed on the pastures of the valleys, and the hillsides. They moved from area to area depending on where they and their livestock could find food and water. Life was on a completely communal basis. There was no private ownership of livestock; some hunted, some guarded the herds, some gathered the berries. In due course this pattern changed and a simple form of agriculture was introduced. By this is meant the tilling of the soil, the planting of seed, and the harvesting of the ensuing crop. This could well be, and still is today in some cases, combined with a nomadic existence based on livestock and hunting. The tribe still moved from one area to another throughout the year, but on a more or less fixed circuit. At the appropriate time it would cultivate and sow, and then move on in search of better pastures, water, or better hunting. But when the crop that it had sown was ready for reaping it would return and gather in the harvest.

Over the centuries the nomadic form of life was abandoned. A spot was chosen where water supplies were plentiful and regular throughout the year, where the soil was fertile, and where living was relatively easy, safe and comfortable. A portion of the land was cleared of bush or forest and the crops were sown and harvested, still on a communal basis. As the needs of the tribe became greater, or as the fertility of the land which had first been cleared was exhausted, more land was cleared and cultivated, and the worn-out clearing abandoned, to be covered again in due course by bush, and thus regenerate its fertility over the years. Still there was nothing approaching the private ownership of land.

Over the centuries, however, the pattern changed: this change took two entirely different forms. One, which is still found today in certain parts of the developing world, consists of a restricted form of private ownership, based on cultivation. The lands of the tribe are owned, by custom or, in some cases, by law, by the entire tribe. The grazing lands are held for the benefit of the tribe, but the livestock may be privately owned, each family of the tribe having the right to a certain number of animals. Similarly each head of family is allocated by the tribal council or by the chief a defined area of land. This he can cultivate as he likes, growing on it what crops he wishes and using whatever methods he may choose.

So long as he remains a member of the tribe, and so long as he cultivates the land, he remains in possession. If he and his family leave the tribe, if he dies and there is no member of his family to succeed him who is capable of cultivating the land, or if he fails for any reason to grow crops on it, the land reverts to the tribe and may be allocated to another member. By custom, it may therefore be said, private ownership on a restricted basis has been established, based on need and ability to make use of the land: but the ultimate ownership of the land still rests with the community.

The other form of change was based on the disappearance of the communal pattern of the society centred on the tribe, with its chief and council of elders, and the substitution of a form of kingship, with the hierarchy spreading down from monarchy through great and petty nobles. This system was normally associated with wars and conquests, and from it grew what in broad terms may be called the feudal system. This can most easily be illustrated by, as it were, a diagrammatic example.

Suppose there are two tribes living not far from each other, both with enough land around them for their own needs, without disturbing the other. Both were originally hunting tribes, and both have now settled and given up the nomadic existence. The first tribe has settled in a fertile valley and has gradually abandoned hunting and become skilled at cultivation. The second tribe has settled in the forest-covered hills, continues to hunt, and only cultivates in small clearings. Gradually its numbers increase and the wild animals become scarcer. It has not the skills, the inclination, nor the suitable land for growing enough food to feed its growing population and to make good the lack of game. It looks with envy on its well-fed neighbours and their green valley. Eventually it makes war on them and conquers them.

It then leaves one of its foremost warriors in charge of the conquered territory, with a few others to help him keep the conquered tribe in subjection. The conquerors still have no taste for farming, but prefer to hunt: the conquered still till the fields, but they must hand over an annual tribute to the victors. Over the years the representative of the victorious tribe assumes the rights of the former chief and council of the vanquished tribe. He may, unless he be very strong, still acknowledge some form of allegiance to his own tribe, and this will be paid by an annual tribute of food and by sending soldiers to help, if necessary, in such wars as the conquering tribe may embark upon. But, with those provisos, he has become king.

If his realm is small he will rule it personally and will claim all the tribute that is payable from those who cultivate the soil. If it is large he will need lieutenants to help him keep order. These he will choose from members of his own tribe or family; or even, as time goes on and tribal origins become blurred, from others with no ties of blood, who commend themselves to him because of their military prowess, their wise counsel or other skills. These lieutenants will themselves collect the tributes that are due, pass on to their overlord his share and retain the rest for themselves.

In mediaeval Europe this system reached a high stage of development. In theory all the land was owned by the king: he kept some of it for himself and granted the rest to the important people of his realm, the bishops, the monasteries and the nobles. In return these were under obligation to provide the king with fighting men when he needed them to maintain order within his own kingdom or to wage war in foreign parts. They might also be required to contribute money or other goods in kind to the royal exchequer. Beneath them were the lesser nobility, holding land on a similar basis, with similar privileges and obligations. So evolved the manorial system, under which the lord of the manor owned the land and kept for his own use the manorial demesne surrounding his own manor house. He kept also certain valuable rights such as hunting and rights of water for water-mills.

The freemen of the manor held, not only for their own lifetime but on an hereditary basis, the rest of the land of the manor. They shared the common grazing land, with clearly defined rights of pasturage and cutting for hay, and they had rights to their own strips of cultivated land, farmed according to strict rules, one year wheat, one year oats or barley, one year fallow. They ploughed the land with their own ox or horse, reaped the corn themselves, and retained the crop, subject to the quantity that had to be paid to the lord of the manor, who also had the right to demand of them a specified number of days' work in the year, and the provision of such things as fish, firewood, or honey.

Strictly speaking the land still belonged to the king: but in fact effective ownership, subject to specified payments and services, was vested in the man who cultivated it. This was the prevailing system in Western Europe. In Central and Eastern Europe the practice was for the noble to own the land under the Crown, but for it to be cultivated on his behalf by serfs, tied to the soil, unable to move elsewhere without his permission, which was rarely granted, under the direction of one of their own number, elevated to the position of overseer or controller. They were paid no wages, but received a certain amount of food and had the right to cut wood for fuel and for the construction of their own houses: in some cases they were allocated a plot of land on which to grow their own food and to keep a pig or some chickens.

As time went on these differences became wider and the effect on agriculture more marked. What may be called the Eastern European system (though it was to be found in other parts of Europe, including southern Italy and also in parts of the Middle East) was transmuted into share-cropping. This term covers a wide variation of arrangements between landowner and cultivator. In some areas, such as the southern states of the United States of America and in many parts of France, it was little different from the landlord-tenant system prevalent in England from the 18th century until the middle of the 20th century, and which will be described in more detail later in this chapter. The share-cropper had a fair degree of freedom to grow such crops as he liked, to cultivate the soil as he saw fit, and to market the produce in any way that seemed to him best. The main difference was that in place of a fixed money rent the landlord received a proportion of the crop. Thus if the crop were poor the landlord suffered as well as the tenant: the tenant was not compelled to hand over to the landlord a sum of money which might represent almost all that he gained from his harvest. In times of prosperity the landlord benefitted as well as the tenant.

In Eastern Europe, southern Italy and the Middle East, on the other hand, the situation was very different. In the first place the landlord usually took a far higher proportion of the crop; in the second place he frequently provided some of the simple implements and draught-power needed for cultivation, as well as the seed. For this he charged a high figure, which was added to his share when the crop was harvested. Thirdly, he often acted as a source of credit to the share-cropper, and this amount, with interest at a high rate, had to be repaid at harvest. The man who actually did the work was therefore left with a very small amount, or even at times a minus quantity, when the accounts were finally made out.

In addition the landowner, through his agent, controlled the actual cropping, telling the share-cropper what to grow and when to grow it, and frequently—as with the 'latifundias' of southern Italy—decided that it was not worth while growing any crop whatsoever for a year or two, leaving the land fallow to regain fertility and waiting for a time of higher prices. The share-cropper was thus left with no income whatsoever from the land during this period. He was, in fact, no different from a farm labourer, except that in place of a regular weekly cash wage he received nothing till the crop was harvested, and then only what was left over after the landowner had taken his share and the amount due for seed and repayment of debts. This system was harsh even when the landowner was in direct contact with the man who tilled the soil. But frequently, in large parts of Asia as well as Europe, the landowner never saw his estate or the people who worked on it. He would spend his life in the city, enjoying himself, spending money and sometimes making it either through court patronage or by commerce. The right to collect the revenues from his estate he would sell on an annual basis to someone else, known as a farmer—a man who for a 'firm' or fixed sum of money acquired from the landowner the right to make as much as he could out of the tenants of the estate; this was the original meaning of the word. Later it came to be applied to the man who paid a rent to the landowner and cultivated the holding.

This 'farmer', either personally or through sub-tenants or paid employees, collected what he could from the cultivators. It was clearly in his interest to extract from them as much as possible, since all that he could get over and above what he had paid the owner was sheer profit. Thus those who grew the crop were ruthlessly ground down, and had neither the means nor the incentive to grow more. This system was especially common in India, under the name of 'zemindary'. As a result the rate of farming progress in the countries where this system of landownership and cultivation prevailed was extremely slow.

In England the system evolved along entirely different lines. In the first place feudal dues and services gave way to cash payments. The freemen of the manor paid their lord sums of money instead of having to carry out tasks for him as payment for the use of the land which they held under him. He for his part paid wages in cash for the services he required. As time went on the Open Field System, whereby all grew the same crops at the same time on land which was virtually held in common, gave way before the Enclosures. These meant that the open fields disappeared, and to each freeman and to the lord of the manor were allocated the manorial lands in proportion to the areas which they had the right to cultivate.

Undoubtedly there was much injustice in the manner in which these enclosures were carried out. Those who stood high in favour with the commissioners charged with the task received the better land, and the worse land went to the others. But from now on the man who farmed the land could use his own initiative and practise his own methods independent of what his neighbours thought should be done. There was thus scope for individual enterprise, and those who adopted something fresh, and were seen to be successful, were in course of time followed by their neighbours, to the ultimate benefit of the whole community.

Under the enclosures the lord of the manor received the larger part of the manorial lands; but he rarely wanted to cultivate all of these himself. Most of them he preferred to rent to the more successful farmers of the village, who thus became his tenants, paying him a cash sum. Over the years strict rules of good husbandry were laid down, which the tenant was bound to observe; but as the average English landlord took a deep interest in the land, and from time to time was in contact with leaders of agricultural thought not only in his own country but in Continental Europe also, he was well placed to be a wise innovator, and stimulated much progress that otherwise would not have taken place. Prominent among such improving landlords of the 18th century were Coke, the great Norfolk landowner, and his neighbour Lord Townshend, the latter being credited with the introduction of the turnip into England, and the former with the four-course rotation of crops—wheat, turnips, barley or oats, and clover—which has formed the basis of good farming in many parts of England until the present time.

Since the landlord did not wish to farm more than his own home farm, and since relations between landlord and tenant were usually close and personal, most tenants could look forward to security of tenure for their own lifetime, and often for their sons after them. This was in marked contrast to conditions, for instance, in France or Central Europe, where the landlord cared little for the country and spent most of his time and all the money he could afford (and often more) in the city, and if possible at Court, where the chances for advancement lay. The English tenant, therefore, had good reason, in the words of the old country saying, 'to farm as if he would live forever', even if he did not adhere to the second part of the exhortation, 'to live as if he would die tomorrow'. The result was great emphasis on the maintenance and improvement of fertility, the breeding of good strains of livestock, and the adoption of improved techniques. This was made all the easier by the partnership which developed between landlord and tenant, the former providing the new ideas obtained by his reading and his contacts with the outside world, as well as the capital needed for the new buildings and drainage demanded by the new systems of farming; and the latter the practical experience and the hard work.

There were of course shortcomings to this landlord-tenant system. Too much power was concentrated in the hands of the landlord. He was free to raise the rent if the farm became more productive as a result of his tenant's efforts and skill. He could bring the tenancy to an end if he fell out with his tenant over politics, religion, or even the preservation of game. Increasingly he was able to draw substantial rents and live a life of luxury without doing anything in return. But in general it worked well, and as a result agriculture and food production advanced rapidly in England from the middle of the 18th century to the latter part of the 19th century.

Only a few miles across the water in France the evolution of landownership was very different. In the 18th century there was no serfdom as there still was in Eastern Europe, but the condition of those who tilled the soil was in many cases little better than was that of the serfs of Russia. Most of the land was owned by nobles whose ambition was to live in style at the court of the king, for which they needed all the money that could be wrung from their estates, which they visited only rarely for hunting or other relaxation. The peasants were rigidly controlled by the owner's agent, even to the extent of not being allowed to spread night soil on their crops for fear of contaminating the flavour of the partridges, not being allowed to cut their hay until the game birds had hatched off, and having to stand helplessly by while the deer and hares and rabbits ate their crops. No matter how great the damage, they were forbidden to kill them. As a result agriculture remained at a very low level, and the opportunities for hearing about new methods, let alone implementing them, were virtually non-existent.

With the French Revolution at the end of the 18th century came a fundamental change in the pattern of ownership. The slogan was 'The land for him who tills it': the great estates, whether of the Church or of the lay nobles, were broken up and the land, including the right to all game on that land, given to the peasants. The effect of this on the standard of cultivation was such that the experienced English agriculturalist Arthur Young, travelling in France shortly afterwards, wrote that 'the magic of ownership turns dirt into gold'.

But although French agriculture made rapid advances at that time, these advances were not maintained into the 20th century. The farms were of necessity small and based on family labour, and, until the introduction of the co-operative movement which was especially valuable for the provision of credit as well as for the marketing of produce, were short of the capital needed for more up-to-date farming. They were consequently at a disadvantage compared with larger units when it came to marketing and processing, especially dairy products and wine. These drawbacks were accentuated by the laws which made it obligatory that a father's estate should be divided equally among all his children. In the case of a peasant holding this meant that what might already have been a farm scarcely large enough to support one family had to be divided among several children. Not only that, but each child had the right to an equal amount of all the land of different quality. Thus a twenty-acre holding consisting of five acres of forest, five acres of vineyard, five acres of arable, and five acres of meadow had to be divided, if there were five children, so that each received one acre of each category.

In course of time this led to extreme fragmentation, with the farmer having to spend more of his time walking from one part of his farm to another than actually tilling the soil. In the days of horses and oxen the waste of time was great: in the days of tractors and combine harvesters it meant that many fields were too small for modern machines to work in. It also meant that the standard of living of the peasant family fell. Sometimes an arrangement was reached between members of the family, whereby only one of their number would remain on the farm, while the rest, although retaining their financial interest in the farm would seek work elsewhere. Even in such cases the life of the peasant became increasingly unattractive when compared with that of the city worker. In spite of strong political pressure (for until recently the farmers of France represented over fifteen per cent of the electorate) which maintained a highly protected market, the standard of living of the farmer continued to decline except in the most fertile areas where farms were larger and modern methods and machinery were employed. This has led to a general acceptance of the fact that modern farming demands larger units than are found in an agriculture based on peasant or family ownership, and the introduction of incentives to encourage the amalgamation of farms into larger units. As a result the agricultural population of France has fallen rapidly in recent years and the average size of holdings increased correspondingly.

In Russia the post-revolutionary period followed a different pattern, as it was to do in other parts of eastern Europe after the Second World War. At first, as in France nearly 150 years earlier, the land was distributed among those who tilled it; but soon the inefficiencies of such small-scale farming were recognised, as was the need for more technical knowledge on the part of the farmer, more large-scale mechanisation, and a greater degree of control by the central government over the food that was produced. Especially at times of food shortages in the cities, such as took place after both World Wars, it was essential for the government to be able to bring food from the countryside, where it was produced, into the towns where the bulk of the population lived. To collect this food from a vast number of peasants, who preferred to eat the food themselves, or to barter it for luxuries on the black market, was an impossible task. It could only be done under a system of large farms operating under a rigid system of state control. So arose the Co-operative and State farms which now account for almost all the food produced in most of the communist countries.

Much has been written about the successes and failures of such schemes, and it is not possible to go into all the arguments, for and against, in this book. A few points, however, should be made. On the credit side, under such a system, the central government can make an overall plan for food production, deciding where the priorities lie, which areas are most suited to growing which crops, how much food can be produced at home and how much must be imported. It can also ensure that at least most of what is produced is collected and sent to whatever area is in need of it, thus ensuring a more or less equitable distribution. It can plan its farms and its production in such a way that the most efficient use is made of machinery, and it can allocate fertilisers according to the needs of the crops and the soil. It can provide each unit with the services of technical experts, so that cattle-breeding and milk-producing farms have their own resident geneticist and veterinary surgeon, arable farms their own specialists in irrigation and in crop cultivation and conservation.

It also offers wide scope for education and promotion, so that the sons of those who work on such a farm have the opportunity to be trained in those aspects of agriculture where their preferences and their abilities lie. They may rise to be the driver of a big tractor, a specialist in agricultural engineering, the man in charge of a team of tractors or of 500 or 1,000 milking cows, a general manager, a marketing expert, or an accountant. Such a prospect is, to most young men, more attractive than that of helping their father, for most of their lives, to milk three or four cows, feed a few pigs and chickens, plough their few acres with a horse, an ox or an old tractor, and, when the father eventually dies, carry on with the same work, now helped by their own sons, unless the sons decides to seek their living in a factory.

On the debit side there is the absence of Arthur Young's 'magic of ownership which turns dirt into gold'. It is held by many that few men, and especially is this true among those who work on the land, give of their best if they are working, not for themselves or for someone they know personally, but for an anonymous absentee boss, whether he be private capitalist or the state, company chairman or commissar. Particularly in the case of livestock, the man who looks after them must consider them as his own, knowing each animal and its idiosyncracies, giving it the personal attention without which it will not give the maximum yield. Similarly the soil has certain idiosyncracies: some fields, or even parts of one field, can only be ploughed when it is dry; with others the ploughman must wait for a shower of rain if he is to be sure of a good seed-bed. Neither man, animal, nor soil will make the greatest contribution of which it is capable if it is no more than a small cog in a very big wheel.

Secondly, there is the remoteness and inefficiency of bureaucracy. The planning done in the offices of the central government may conform to the beliefs of the professors of soil science and animal nutrition, may take full heed of the soil surveys and meteorological records, as well as the overall national requirements: but by the time they have been broken down to regional and local level, and have passed through the hands, and the files, of lesser bureaucrats, the instructions that eventually reach the farms themselves may be far removed from the realities of farming. Yet those who have to do the work must do their best to carry out the instructions they receive, no matter how much they conflict with the experience of practical men.

Thirdly, there is the problem of innovation. Many of the most important advances in agricultural practices have come about through the original thought of one individual who has had sufficient confidence in himself to put his ideas into practice with his own money, or that of a few others whom he has persuaded that the new scheme is likely to succeed. Often such new ideas fail, but sometimes they succeed and have a big impact on farming methods, even though they have been rejected by the established experts. With a highly centralised system such forms of progress are not possible. Before a new idea can be tried out the appropriate heads of government departments and research institutes have to be convinced that it has some weight: because of the bureaucratic system it is rare for them to be approached direct: usually it must be through intermediaries and subordinates who may not have the same open-ness of mind that good scientists should have. Regrettably this is sometimes absent even among those who reach the highest ranks in government service.

At this stage one can say no more than that in some countries which have adopted the system of state or collective farms as opposed to private ownership by the farmer himself, or a system based on the landlord-tenant relationship, progress has been markedly faster than in the old days of the previous regime. In some it has been distinctly slower than the progress that has been made in countries working other systems.

It must also be recognised that the Soviet Union, which was the pioneer in this form of farming and landownership, has markedly failed to produce enough food for its own needs, and has to make massive purchases of grain from capitalist countries in order to supplement the shortfall from its own farms. But there can be no doubt whatsoever that the form of landownership will have a profound effect upon the rate of agricultural progress, and therefore must be studied with care by those who are concerned to increase food production.

Each country will have to work out the system that suits it best, and it is probable that more than one system will be adopted. Where there is a strong tradition of communal ownership a collective or co-operative unit may be best. In others, private ownership may be preferred. In yet others, the land may be owned by the state while it is cultivated by individual tenants. In such cases the tenants should have complete security of tenure provided the land is put to good use. Furthermore they should have the privilege of passing the farm to one of their children, or even the whole family, provided it is not split up to such an extent as to make it more difficult to farm efficiently.

Wherever there is independent occupancy there must be co-operative or governmental organisations to provide the necessary services, whether of machinery, buying and selling, credit, or technical advice. Where there is no communal tradition, but the farmer has been the tenant or share-cropper of a landowner, unless the landowner is in a position to fulfil properly his obligations as the provider of long-term capital for buildings, drainage and irrigation, it might be well for the state to take over the land and become the landlord. No useful purpose, however, would be served by such a change unless the state had a well-developed group of people experienced in the functions of a landlord, and sufficiently independent of centralised control to be able to act on their own initiative. They would also need sufficient funds to enable them to do what was necessary, but with supervision adequate to prevent favoritism and corruption from creeping in.

Whatever may be the future of existing private landlords they must not be allowed to take an undue share of the profits arising from the land while doing nothing in return. This is not to say that the farmer should get his land at a low price. As with any other commodity, whether it be fertiliser, feeding stuffs, machinery, and above all labour, if it commands too low a price it will not be put to its full use. Within reason, the more money a farmer has to pay for his land the more he will extract from it. If he gets it cheap he will leave parts of it uncultivated, will not drain the damp corners, and in other ways fail to cultivate it to the full. The national asset of the land will be wasted.

Where whole new areas are brought into cultivation as a result of large-scale irrigation or drainage schemes, or by opening up land by the construction of new roads and airfields, and even of towns and villages, direct state farming on a big scale may be preferable, or private companies with experience of such forms of farming may be encouraged to come in and buy the land or, better still, to take long leases or enter into joint enterprises with the government. Alternatively, if skilled farmers are available from elsewhere in the country, where pressure on existing supplies of land are great, they may be encouraged to settle in the new areas and given special help and incentives to encourage them to do so. A fine example of such reclamation and settlement is found in Holland, where fertile land has been recovered from the sea and farmers from other parts of Holland have been brought in to cultivate this new land.

In deciding which of these systems to adopt, and what particular mixture of different systems should be followed, two further points must be borne in mind. The first is that while there may be theoretically optimum sizes of farm for different operations, based on the proper utilisation of the equipment that is needed for efficient farming, the success or failure of a farm must in the final analysis depend upon the skill of the farmer. Many farmers have the skill to manage a farm of 50 acres or a milking herd of 20 cows: some can cope with 500 acres or 50 cows: a few will make a success of 2,000 acres or 150 cows: only a handful are capable of dealing with more. Whatever the theoretically optimum size may be, in practical terms the size of farms must be geared to the number of farmers capable of running a farm of a given size successfully.

It will therefore be found that the correct size of the average farm is invariably smaller, at the outset, than the theoretical figures indicate; but as time goes on, and as existing farmers and their sons gain experience, the size can be increased. There must therefore always be provision to expand the size of farms as the years go by, either by bringing more land into cultivation, or by the amalgamation of existing holdings. An alternative to extending the actual area of the farm is to increase its productive capacity, whether by irrigation, by the introduction of more intensive cropping, or by adding to it, for instance, a livestock unit.

The second point concerns the need in every district for at least one farm where neighbouring farmers can see new techniques being put into practice. No matter how down-to-earth and how persuasive the advisory officer may be, his task will be made ten times easier if he can take the farmers of his district to a farm not dissimilar from theirs where they can see the methods which he has been trying to persuade them to adopt actually being carried out with success. Such farms may be run by the state, by the nearest university or farm institute, by the Agricultural Co-operative Society, or even by an ordinary commercial farmer selected by the advisory officer because of his skill and willingness to try new methods. Whatever form they may take they will play an invaluable part in helping to spread the discoveries of the scientists more rapidly throughout the farming community.

XII
Waste

Mention has already been made of the waste of some of the major ingredients needed for food production. Valuable chemical elements and humus which are found in sewerage, factory waste and other effluents are not used: water, which even in dry parts of the world may fall in heavy storms at certain times of the year, is allowed to rush down otherwise dry gulleys and into water courses and rivers, and thence into the sea, instead of being trapped in dams and conserved for the dry season: and this same water, so valuable at certain times of the year, is itself a formidable source of destruction by carrying with it, when it falls in tropical downpours, much of the topsoil, leaving behind nothing but bare rock. Similar waste of soil by erosion takes place even in areas of flat land through the action of wind if good farming methods are not followed, or in the absence of protective forests.

The subject of this chapter, however, is the waste of the crop itself which occurs from the time the crop starts to grow until the time when it is finally consumed. Such waste can take place at three stages: first, between the planting of the crop and its harvesting; second, between harvesting and its appearance in the shop or market in the form in which it will finally be consumed; and third, from shop or market to the table.

The commonest cause of waste in the actual growing of a crop is disease. The yield of a promising field of wheat can be halved by an attack of rust, of sugar beet by virus yellows. Maize, soya beans, rice all have a variety of insects, fungi, and viruses which can play havoc with a promising harvest. Black pod and swollen shoot can ruin an entire crop of cocoa. Scientists have provided the cure for or protection against most of these dangers, and whenever a new disease appears they quickly set to work to discover how to combat it. Ignorance on the part of the farmer is one of the reasons why those diseases for which a protection is already known is not used: but that is only part of the story. Cost is another.

In most cases the cost of the protective material is high, and in some cases the cost of application is high also, especially if the material has to be sprayed by aeroplane. Unless the value of the harvested crop is going to be high, it simply is not worth the farmer's while to spray, especially as, even after treatment, the disease may continue to cause damage. On the other hand, even if no treatment is given the weather may suddenly change and the disease disappear.

It is true, for instance, that an attack of rust may halve the final yield of a crop of wheat. But when the disease first appears, and that is the time when the spray must be applied if it is to be effective, no one can tell how much damage it will do. A full crop may be worth £150 per acre, and the cost of the spray and its application may amount to £7·50 per acre. In such a case the farmer will probably decide that such an expenditure as an insurance premium against the loss of anything up to £75 is worth while. But if the value of the crop is unlikely to exceed £75 per acre, and the cost of treatment is £15, it would still be worth while if he were certain that without the treatment he would lose half the crop, or £37·50: if, however, the weather changed, the disease disappeared, and the crop grew away well, he would have paid out £15 per acre for little if any benefit.

It therefore comes down to an assessment of the risk of the disease spreading or disappearing, and, above all, a balance between the cost of the treatment and the value of the crop. The more valuable the crop the more likely the farmer is to spray. The more money he receives, the less will be the avoidable waste.

Another source of waste is the damage done by rats and other rodents, and by birds and insects. Until the middle of this century the most dramatic, and probably the greatest cause of this type of damage in tropical countries was the locust. Hundreds of acres of a crop just ready for harvesting would be stripped bare in a matter of hours by these grasshopper-like animals, and no effort on the part of the farmer could prevent it. Today, thanks to the Locust Control Service and the use of insecticides sprayed by aeroplanes, such widespread devastation no longer takes place. But for all that the locust is still a serious cause of waste in the growing crop.

Birds are particularly inclined to ravage orchards and vineyards when the fruit is ripe, and also to destroy substantial areas of arable crops such as peas and wheat in temperate climates. To prevent such damage, apart from the wholesale destruction of birds which is both difficult and unacceptable on environmental and conservation grounds, manpower is needed to scare the birds away. When wages are low and manpower is plentiful, people, especially children, can be sent into the fields, orchards, and vineyards from dawn till dusk to scare the birds away. With higher standards of living this method becomes more difficult and more expensive, so mechanical devices have to be used instead.

The damage caused by rats to growing crops is more serious than that caused by birds. Especially is this so among the tree crops producing vegetable oils, such as oil palms and coconut palms. Here poison, mixed with grain as a bait, is the most effective means of control, and although the cost is high it is always worth while. But even where control is efficiently carried out it may well be that ten per cent of the crop is lost through rat damage. The more widespread the control of rats is throughout a district, the more effective. If only some of the farmers do it, the rats will breed in the adjoining areas and move into those parts where the rat population is lowest because of good control. It is therefore something which, especially in regions of small farms, should be carried out on a district basis and not left to individual farmers to control or not as they see fit.

Even when the crop has come to maturity and is ready for harvesting there can be waste. This is caused mainly by the hazards of weather. A hailstorm when the grapes are ripe, high winds when the grain is fit for the combine harvester, a prolonged period of rain when the sugar cane is mature, all result in lower yields. Both with sugar cane and with sugar beet there is a relatively short period when the sugar content is at its highest: if harvested before or afterwards a smaller yield of sugar is obtained. Therefore to get the maximum yield harvesting should be carried out during the few weeks when the sugar yield is highest. Similarly, because the sugar content of both cane and beet falls after harvesting and during storage, processing should take place as soon as possible after harvesting. The cost of harvesting machinery, however, is high, and the cost of processing equipment is higher still. It is therefore desirable to spread the use of both machinery and plant over as long a period as possible.

So a balance has to be struck between these conflicting interests. The cheaper the final product relative to the cost of machinery the more economic it is to have a long harvest period, even at the cost of losing a relatively high proportion of the crop. If the value of the crop rises relative to the cost of machinery, so it becomes economic to increase the investment in machinery, and harvest and process a larger proportion of the crop at a time when the yield is highest. This is yet another example of how the waste of a crop that has actually been grown can be cut down by raising its price to the producer.

A further waste of potential food arises directly from low world prices. In the United Kingdom alone it is estimated that in recent years straw is burnt which, if suitably processed, would produce animal fodder equivalent to 300,000 tons of grain. The tops of sugar beet are ploughed in, and these, if collected and ensiled, would produce animal fodder equivalent to 500,000 tons of grain. With the low prices of fodder grains that obtained until 1974 neither of these operations was worth while. It was cheaper for the farmer to burn his straw, plough in his beet tops, and buy feed grain, thus destroying in the United Kingdom alone the equivalent of more than half a million tons of grain. Similar waste takes place in many other rich countries, so the amount of potential fodder destroyed in this way must amount to many millions of tons.

Waste on the farm does not occur solely among crops. It is to be found also among livestock. The most obvious cause of waste is disease. As with crops there is already much knowledge concerning the control of disease. Although more research is essential, the greatest need is for the fullest possible use to be made of existing knowledge. This means that there must be more qualified veterinary surgeons and that they must be in direct contact with the farmers who actually look after animals. But disease is caused not only by the presence of infective organisms: it flourishes under conditions of poor nutrition and may also be caused, or accentuated, by unhygienic living conditions and by hereditary factors. Good advice can help to overcome all these causes, but in most cases money is needed too, money to pay for a good and healthy bull, or for the use of artificial insemination, and money to replace old and unhygienic buildings with modern hygienic ones.

There is one sphere in certain developing countries where much can be done to overcome waste in livestock production without any expenditure of money whatsoever. Especially in areas which concentrate on livestock, it has for long been accepted that a man's wealth is reckoned according to the number of cattle he owns. It matters not if the cattle be sickly or healthy, thin or fat, give much milk or none at all: the man who owns fifty head of cattle is richer than the one who owns only twenty-five. He is therefore just as loth to dispose of any of his livestock as is the owner of an ancestral mansion in Western Europe loth to dispose of his home and move into a smaller one when he cannot afford to repair the roof of the big house, or heat or furnish it properly, and when the smaller house would give him more convenient and more comfortable living and allow him more money to spend on other things.

Such an owner of cattle must be persuaded that it is wasteful to keep more than he can feed properly. He must be shown that the wise thing to do is to reduce the number of his herd so that the cows will breed more regularly and give more milk and the calves will have sufficient food to grow and fatten quickly. The same lesson must be learnt by those who keep sheep, pigs and chickens. A small number of animals looked after well and fed well will produce more wealth for the owner, and more food for others, than will a larger number of half-starved animals. By reducing the numbers to the amount that the pastures and the available fodder can support, a large degree of waste will be avoided.

After the crop has been harvested and before it appears in the shop or market there are further opportunities for waste. In the Bawden Memorial Lecture of 1974 Dr H.C. Pereira, Chief Scientist at the Ministry of Agriculture, Fisheries and Food of the United Kingdom said: 'The estimate of several sampling surveys carried out by food scientists shows that the nation as a whole wastes some twenty-five per cent of the food supplied from both home-grown and imported sources'. This figure was arrived at by calculating the total calories in all food arriving at United Kingdom ports, and all food sold off United Kingdom farms, making allowance for the known proportion of inedible matter, and comparing this with the figures for calory consumption per head of the population as shown by dietary surveys.

In Brazil it is calculated that from five per cent to twenty-five per cent of the maize crop is lost annually because of deficient storage and transport facilities: and nineteen per cent of the total agricultural product is lost during distribution. In 1973–74 Ecuador had more than one million tons of bananas for which no market could be found, and most of these were not harvested. In the case of grain crops, especially those grown on a small scale, waste can take place in storage on the farm, principally from rats and mice but also from weevils and similar pests. In temperate climates if the grain crop is not harvested by combine harvester it will be stored in stacks, under cover or in the open. If the latter, there is always the risk of waste from rain which can partially be avoided by thatching with straw from the previous year's crop. This is effective for a short while, but thereafter birds make holes in the thatch and the rain or snow penetrates. Even when adequately protected from the weather, rodents will enter the stack and eat or damage a significant proportion of the grain.

If the crop has been combined, as is becoming increasingly the general practise, there can be complete protection from the weather, and from rodents also, provided the threshed grain is stored in metal, brick or concrete silos, though there can be a loss of anything up to five per cent of the crop in the actual operation of combining, unless the operator is very skilled. But if the storage is on the floor in a barn there is always the risk of rodents, which, as with field crops, must be controlled by poison. There is also the risk, when combine harvesters are used, that the grain that is stored in bulk heats, because of excess moisture content at the time of harvesting. It requires skill on the part of the farmer to ensure proper movement and ventilation of the grain, and perhaps the installation of an expensive drying plant, burning expensive fuel, in order to reduce the moisture to a level at which it can safely be stored. Such plants are usually too expensive for the small farmer, so there is need here for central drying and storage installations operated on a co-operative basis, commercially by private merchants, or by central buying and marketing organisations. Without such provisions post-harvest and pre-sale wastage can be considerable.

Apart from such risks, the storage, transit and processing of non-perishable crops present few problems, although there is always some danger of damage from sea or weather during transit or while waiting for shipment or on the dockside at the point of unloading. The same does not hold good for perishable crops. With fruits and vegetables the losses are heavy. First there must be selection on the farm itself; all that is manifestly of inferior quality because of size or damage while growing or during harvesting must be rejected. Large quantities of such crops as potatoes, tomatoes, apples, citrus fruits, bananas, and cabbages are rejected at this stage. Some of these may be fed to pigs, but because the supply is variable and often seasonal it is impossible to ensure that there are enough pigs at the right time to eat all that is rejected. In cases when the supply of rejects is large and fairly regular, at least for several months on end, it may be possible to process them into animal feed by dehydration, but this can only be economic in rare cases and on a large scale. It also presupposes a high price for alternative supplies of fodder. The main attack on this form of waste lies in the breeding of varieties that produce the largest proportion of product of even quality, and in methods of cultivation which minimise damage during growth and harvesting.

When such perishable and usually delicate crops have been despatched from the farm itself either to market or to the point of further selection and packing, there are yet more chances of waste, both through more rigid selection and because of damage in transit; this risk of damage continues right up to the time of arrival at the retail outlet. Waste here must be reduced by improved methods of packing, stronger containers and more careful handling. The third of these depends on the degree of supervision exercised by those who undertake the transport; the first on greater skill on the part of those who actually do the packing, as well as better supervision; and the second partly by more study of the best methods of packaging, but also, as is so often the case in the elimination of waste, by an alteration in the cost of the packaging material relative to the value of its content. If the latter rises faster than the former it will pay to use a more expensive form of packaging.

Many crops, both perishable and more long-lasting, are sent from the farm to factories for processing. A large quantity of meat and poultry is treated in this way, especially since quick freezing has become common, as are fruits and vegetables and milk. Provided the processing plants are efficiently run, as most of them are, there is little waste here once the raw material has passed the requisite tests for quality, purity and hygiene. But some of it, especially meat and to a lesser extent milk, fail to reach these standards. The animals may be found to be suffering from disease which renders them unfit for human consumption, or the milk may come from diseased cows and the infection is transmitted in the milk, or it may have become contaminated between leaving the cow and arriving at the factory.

Here the control must be exercised on the farm itself, and in general the processing factory is the best vehicle for ensuring that this is carried out, especially if the meat or the milk is produced on contract for the factory. This must be done in collaboration with the veterinary services and the department responsible for public health; but it must be done in such a way that the advice given is both readily understandable by the farmer and easily put into practice. If, as may often be the case, expenditure on new buildings or equipment is needed, credit must be available from the factory with which the contract has been made, or from one of the other sources described in a previous chapter.

Then there is the form of waste which results from the inability of the farmer to find a market for his crop. Mention has already been made of the state of affairs between the wars, when coffee was burnt and wheat thrown away because of lack of a market. On a smaller but significant scale there are still examples of farmers finding that they lose less money by allowing crops to rot on the ground rather than incurring the expense of harvesting them and sending them to market. No matter how perfect the planning of production and the organisation of markets, there will always be periods when certain areas have a surplus of a given perishable crop, above the immediate demands of the consumer and above the capacity of the processing plants. Even though there will undoubtedly be people in other parts of the world who would dearly love to have such food, the cost of handling and transport would be insupportably high. Such wasted crops must be kept to an absolute minimum, and this can be done by planned production and distribution, and the financial provisions described in earlier chapters. Without these, this form of waste will inevitably continue.

Finally there is the waste that takes place once the food has reached shop or market. Here it is impossible to generalise or even to form any firm impression of the amount of waste that takes place. In poor countries it is unlikely that any great quantity of food is wasted in the household itself, though frequently ignorance and custom in cooking methods can lead to substantial waste of nutritive value in the food itself. Waste must also inevitably take place with home-produced perishable food at certain seasons, when more is produced than can be consumed and there are no means to preserve the surplus, or knowledge of how to do so. In the cities the advent of refrigeration has reduced waste greatly; but there is still a significant amount in the retail sector, especially in the smaller shops.

In the rich countries the picture is different. In all but the very poorest households waste takes place daily, and in the middle and upper income brackets this waste can be significant. It is probable that little animal protein is directly wasted, though full use is rarely made of bones and fat. In expensive restaurants on the other hand there is conspicuous waste of meat due to individual portions being too large for the appetities of the customer and being left on the plate and then thrown away. There is also significant waste in canteens, hospitals and schools, but milk, eggs, and sugar are rarely thrown away. With fruit and vegetables waste is far greater; much is thrown away because of loss of freshness, but there is waste also in the preparation, when much edible material is discarded before coming to the table. It is with these perishable commodities that there is inevitably waste in the shops themselves when they fail to sell out before deterioration has gone too far.

The greatest avoidable waste occurs with bread. People in rich countries prefer to eat their bread fresh and therefore are in the habit of buying it daily, or every other day. It is a relatively cheap food, so they have no hesitation in buying a bigger loaf than they will need, and indeed few bakers bake a loaf that is small enough for a small household that eats no more than a slice of toast for breakfast. Every day, therefore, large quantities of bread are thrown out, carted away with other refuse, and disposed of with the rest of the garbage. While there are no statistics to show the total amount of bread wasted in this way, I was told by a high official in the United States Department of Agriculture shortly after the war, when bread was strictly rationed in Germany, that the amount of bread thrown away daily in the city of New York alone was enough to make bread rationing in Germany unnecessary.

Little can be done to prevent household waste of this kind, although the more expensive food becomes the more care the housewife will take. However, the separation of food waste from other refuse, and its processing, as was carried out by some municipalities in Britain during the Second World War, could result in the production of worthwhile quantities of animal feeding stuffs which would to some extent mitigate this waste.

While not in the same category as food waste, much food which could feed human beings is diverted to other uses, and much is used for non-food purposes. For instance, every year in the United States of America $4,000 million are spent on pet foods, and in Britain the figure is £180 million. Some of this is produced from materials which are not fit for human consumption, but some, after suitable processing, could be used as human food with no risk to health; much of it has a grain base, which could either be used for direct human consumption or fed to animals which themselves would provide human food.

In the United Kingdom 100,000 tons of fertiliser are spread every year on golf courses, cricket fields and ornamental lawns and gardens. When food production in many parts of the world is held back by lack of fertiliser such use cannot be looked upon as anything but a waste of resources.

There can be no doubt that between the planning of a crop and its final consumption as human food much waste takes place. Some of this is due to ignorance, some to lack of care, and some to natural causes. It can be lessened by more research, more education, and by propaganda in the developed countries directed towards reducing waste by consumers. It will also be reduced at all stages if the price of food were to rise.

It is tempting to try to arrive at some estimate of the additional amount of food that would be available throughout the world if all waste were eliminated. Nothing approaching an accurate figure can be arrived at, but the following calculations give some idea of magnitude of savings that could be effected.

At least 10 per cent, and in many cases more than 20 per cent, of the food grown is wasted on the farm. Let us say, therefore, conservatively, that with no waste at this stage food supplies throughout the world could be increased by 15 per cent. Between farm and table we can take Dr Pereira's figure of 25 per cent wastage for the rich countries—say one quarter of the world's population—and 10 per cent for the remaining three-quarters, making an average of nearly 14 per cent. If only half of this total wastage were saved, and if all this saving were made available to that half of the world's population that suffered most from malnutrition, it would be possible to increase their diet by 30 per cent.

The possibility of increasing the availability of food by something of this order, without having to grow an ounce more, without having to use a single extra man or machine, without bringing into cultivation a single extra acre, must surely justify much study.

XIII

Some Who Have Tried

During the second half of the 20th century great strides have been made in most developing countries towards improving their economic position, helped by money and technical assistance from the developed countries. It is not surprising that by far the greater part of these efforts have been directed towards industrialisation and urbanisation, for the poor countries have looked towards the rich ones and have seen that their wealth appears to stem from industry, commerce, and banking, while their own poverty is associated with a predominantly agricultural and rural way of life.

There have been some, however, who have made great efforts to improve their own agriculture, and with it the quality of life in the countryside. The greatest country that has done this is China; and it would be of interest to make a serious study of how the Chinese Government has set about this task and what measure of success it has achieved. But this is a task which warrants a whole volume on its own; furthermore, statistics for the pre-revolutionary situation in China and comparable ones for recent years are impossible to come by. Any description, therefore, of the means adopted in China, the amount of resources devoted to the countryside, and an assessment of the results, would be both very superficial and probably misleading.

Instead, brief accounts will be given of what has been attempted, and such results as have already been achieved, in four other countries, two large and two small. No attempt is made in these descriptions, to go into detail or to give anything more than a very tentative assessment of results. Furthermore, no judgment is made as to the forms of government in these countries, their superiority or otherwise over previous administrations. The reason for writing of them is to give an idea of the direction in which some countries are trying to move, the priorities they have laid down for themselves, and, where possible, the successes or failures they have met with in the relatively restricted field of general rural improvement.

The countries selected have different histories, different social and economic backgrounds, different soil, and different forms of government. In alphabetical order they are Brazil, Cuba, Iran and Israel.

Brazil is a huge country with a vast reserve of potentially fertile soil, a climate ranging from tropical to temperate, a large population of 102 million but a very low population density of only about 31 persons per square mile, a small proportion of rich and well-educated people, and a high proportion of poor illiterates.

Cuba is a large island in the Caribbean, covering about 70,000 square miles, with a population of about nine million and a population density of 130 people per square mile. Before the revolution in 1959 it had a small number of rich people, living mostly in the capital, Havana, very little industry, a flourishing tourist trade, mainly with the United States of America, and an economy depending partly on tobacco but to an overwhelming extent on sugar, which was in the hands of rich individuals and large companies, with a substantial proportion in foreign ownership. Much of its food comes from overseas, paid for largely by sugar exports, with substantial price support before 1959 from the USA, and since then from the USSR.

Iran covers 636,000 square miles, with a population of 32 million, giving a population density of 50 people per square mile. It has large regions of fertile soil, but suffers from lack of well-spaced rainfall, which in the past has been to some extent counteracted by irrigation systems stretching back over more than a thousand years. It has huge deposits of oil, and other mineral resources too, which now bring in enormous quantities of foreign earnings. Until recent reforms, its form of landowning and cultivation has been based on very old traditional patterns. As in Brazil and Cuba, the distribution of wealth has historically been very uneven, as has been the availability of education, with a consequent high rate of illiteracy, especially in the rural areas.

Israel as a state has existed only since 1947. It is a small country of only 7,990 square miles and a population of 3,200,000, giving a density of 400 people per square mile. As in many parts of Iran, it has a considerable proportion of potentially productive land, but suffers from drought. Before the state of Israel was established the land formed part of the Palestine mandate, administered by Britain. The population was small, and, although there were areas of good farming, food production was incapable of feeding more than relatively few people at a low standard, and conditions of life, especially in the countryside, were poor.

About the middle of this century the Brazilian government took the major decision to encourage people to move from the prosperous coastal strip close to the capital of Rio de Janiero and the even more prosperous industrial centre of Sao Paulo into the interior of the country, which was underpopulated in the extreme. The first step towards implementing this policy was the construction, in the early 1960s, of the new capital, Brasilia, some 1,000 km into the interior, in what had been, till then, uncultivated bush.

Since then further steps have been taken to open up the vast country. By 1974 there was a 3,700 km highway running from north to south, and another, from Belem at the mouth of the Amazon, to Brasilia, a distance of 2,000 km. Two more, each of 3,200 km have been made in the north and in Amazonia. Between 1969 and 1973 the country's road network grew 20 per cent from 1·1 million km to 1·3 million km, with plans for a network of 20,000 km in the uninhabited Amazonian region.

At the same time a 10 km belt of land on each side of the Trans-Amazon Highway was expropriated and made available for settlers in order to cultivate the most suitable soil along the route. These settlers are helped both financially and technically, and are housed in a series of settlements consisting of approximately 50 families. By 1974 3,000 families had been settled in this way: the target is 100,000 families before 1980. The settlements are grouped round a larger centre, known as an agropolis, in which there are health services and schools, technical services, banks, and the headquarters of the rural co-operative.

In the educational field extensive use is being made of radio. Since 1970 a supplementary education service under the supervision of the Ministry of Education has been provided, directed especially towards the rural population. In 1970 59 per cent of rural families owned radios; by 1972 the figure had risen to 85 per cent. The educational programmes for adults take place outside normal working hours. By 1978 there will be educational television programmes specially for the rural areas of the west of Brazil. In 1971, a Programme for the Special Assistance of Rural Workers was instituted. This provided for disability and old-age pensioners, as well as medical services, preventive medicine, and rural sanitation.

In 1967 the Brazilian Institute for Land Reform carried out a survey that showed that 25 per cent of the land was owned by 0·1 per cent of the largest landowners, while 76 per cent of the smallest landowners owned only 14 per cent. Five per cent of the total area of existing rural holdings were owned by only one hundred people. Since then a policy of land reform has been carried out, its objects being to increase the efficient use of land and create a more balanced pattern of landownership. In the three years from 1971 to 1974 50,000 new holdings have been created in this way. Much of the land has been expropriated on a voluntary basis, the owners receiving not only payment for the land but credits to help with the development of the land which they retain.

Brazil has now become a major manufacturer of tractors and of fertiliser. Between 1968 and 1973 the number of farm tractors increased by 500 per cent to 250,000, and, in the same period fertiliser consumption rose by 300 per cent. The plans for the future are even more ambitious: by 1980 it is hoped that one million hectares will be under irrigation, and that rural electrification will have increased from the 1974 figure of 11,000 km to 114,000 km.

It will be seen that the greater part of this programme for rural development did not start until the 1970s, and that the greatest effort is still in the planning stage. Between 1960 and 1970 agricultural production expanded at an average rate of 4·4 per cent per annum. In 1971 this figure rose to 11·4 per cent, but fell back to 4·1 per cent in 1972 and less than 4 per cent in 1973, due mainly to climatic conditions, thus giving an annual average increase for the three years of some 6 per cent. It is intended that the average rate of increase between 1970 and 1980 should be 7 per cent. Some of this increase will be exported: this applies especially to coffee, the most important crop in Brazilian agriculture; but also to sugar, vegetable oils, and meat. The rest will be used to improve the standard of living of the population of Brazil, now growing at the rate of 2·5 per cent per per annum.

In Cuba economic policy since the revolution has been directed largely towards the development of its agriculture. Having little new land to be brought into cultivation it has concentrated on intensification and diversification, while still retaining sugar as the main crop; on mechanisation, irrigation, and the increased use of fertilisers; and above all on improvements in the conditions of life and the educational facilities of those living in the rural areas.

The first target of the diversification programme was the development of a livestock industry, based on milk, and concurrent with this the creation of fresh pastures and the improvement of existing pastures. The second target was the creation of a substantial citrus industry based on exports, as well as the expansion of coffee production. At the same time it was hoped at least to maintain, and if possible increase, sugar production.

After fifteen years of this policy sugar production is at a level of about 8 million tons (these figures are not made available by the Cuban government, so can only be an estimate), compared with 5·78 million tons in 1958, and 5·96 million tons in 1959.

Before 1958 the dairy industry was small, and few of the poorer people had milk. By the early 1970s nearly a million cows had been inseminated yearly with semen of high-yielding imported dairy bulls, with the result that by 1970 690,000 tons of milk were produced annually: by 1974 this had risen to 799,000 tons, and there was an allocation of one litre of free milk daily for every child under one year old, as well as supplies of milk and other foods for school children and children of pre-school age.

One of the results of this has been a marked decline in infant mortality, and the virtual disappearance of tuberculosis among children, as well as a general improvement in health. Thus the mortality rate for gastro-enteritis in children under one year fell from a figure of 134·6 per 10,000 live births in 1962 to 32 per 10,000 in 1973. So far as the population as a whole is concerned there has been a marked improvement in the level of nutrition. The consumption of milk per head rose between 1963 and 1973 by 26 per cent, of fish by 43 per cent, and of cereals by 19·5 per cent.

The citrus industry has also developed significantly. In 1974 the area under citrus was six times what it was in 1952, and it is planned that by 1980 production will be 550,000 tons compared with the figure of 187,000 tons in 1974.

Conditions of life in the countryside have been improved by massive investment in roads, schools, hospitals, and housing. Between 1959 and 1974 16,970 km of new roads of all kinds have been built, which is 1·7 times the length of roads in 1959. Between 1970 and 1974 road building was taking place at an annual average of 545 km of motorway and highways, and 1,453 km of secondary roads, while between 1976 and 1980 it is planned to build 1,000 km of motorways, 4,000 km of highways, and 7,000 km of secondary roads.

So far as education is concerned, the expenditure on school building doubled between 1965 and 1973, and the great bulk of this expenditure went to rural areas. In 1958–59 there were 7,588 schools in Cuba and by 1972–73 there were 15,972. During the same period the number of pupils rose from 780,000 to 2,075,000.

All the new secondary schools in Cuba are boarding schools, and these have been built in the countryside. The pupils are responsible for some agricultural enterprise, usually a citrus plantation; thus they are taught as part of their general education not only some of the practical aspects of farming but are also brought up to realise its significance in the national economy. Because all such schools are in the country there is no incentive for parents to move to the town in order to ensure for their children the best possible education. Similarly most new clinics and hospitals have been built in the country, and the first drive for improved housing has taken place, not in the slums of Havana, but in the country districts, where new villages have been built, with blocks of simple modern flats, shops and community centres as well as schools and clinics. Main water and electricity are now available in almost every rural area.

This concentration on agriculture is found also in higher education, where together with engineering, it ranks high among the subjects that are taught there, while the law accounts for only a very small proportion of the students. The same priorities also hold true on the material side, where, with a communist system of state-regulated prices and wages, it is possible to control the financial rewards reaped from different activities. Those who are engaged in agriculture, at all levels, are favourably placed compared with those at similar levels in other occupations. While these benefits are particularly apparent for those who work on the large state farms, which account for by far the largest proportion of agricultural output, the 200,000 or so small private farmers, principally engaged in tobacco production, reap benefits also, not only from stable prices but from the amenities that have been brought to the countryside.

In Iran what has come to be known as the White Revolution started in 1963. Up to that time the country was still essentially feudal. The large landowners, mainly absentee, formed the ruling class. They owned most of the arable land, which was cultivated by peasants whose conditions were not very different from those of serfs; and cultivation methods had changed little over the centuries. Industrialisation was against the interests of the landowners, who wanted large supplies of cheap labour without any competition from factories; and for similar reasons they were opposed to the spread of education in the rural areas. The average rural family had a per capita income of between $2 and $7 a month in 1963, and the total national income was about $3 billion, or $130 per head per year. By 1973 this had risen by 350 per cent to $500 per head per year.

Before 1973 agriculture took about 75 per cent of the Iranian manpower but contributed less than 30 per cent to the gross national product, oil accounting for 17 per cent. Villages were remote communities, cut off from the rest of the country, and enjoyed virtually no health or educational services. Naturally all those who had the chance left the villages for the towns, where the opportunities for a better life were infinitely greater. The Shah himself wrote in 1961 (*Mission for my Country*) '... A small number of big landlords—no more than a few dozen—own from several to forty or more villages each, and some of the largest are tribal chieftains. Much of the worst-managed land in Iran is in the hands of the biggest landlords. Typically they are absentees who give little thought to improvement or the welfare of the peasants and entrust management to professional overseers who may ruthlessly exercise their power over the peasants while the landlords enjoy themselves in Tehran or Europe or America. It is true that among our biggest land-holders are exceptional men who display notable social consciousness; but as a class the big private landlords are parasites ...'.

The Shah had already in 1950 distributed 2,000 villages, estates, and farms of the Crown Estates to those who actually cultivated the land. By 1962 a total of 200,000 hectares had been distributed among the 42,000 farmers who had been working on them, and in 1952 a Development Bank was set up to provide the money needed for machinery, irrigation and farm houses. In 1962 a law was passed restricting ownership to not more than one village; and in 1964 the second phase of land reform began, aimed at abolishing the old landlord-peasant relationship and creating a new class of small owner-occupiers. By 1972 over 17,000 villages and large estates had been bought from the previous owners and over 750,000 farmers had obtained land of their own. In many cases, they also required the rights to irrigation water which, in the Iranian climate, is in some regions of equal importance to land.

The second phase of land reform consisted of restricting the amount of land that could be held by an owner who did not farm the land himself, the maximum differing in different areas, depending on soil conditions and availability of water. A considerable amount of freedom was given to landowners affected by this legislation to make such arrangements as they liked with their former tenants. By 1971 54,000 villages and 21,000 farms had been distributed under this legislation, and nearly 2½ million farmers had received land.

At the same time as those who actually cultivated the soil were becoming independent of the former landlords, encouragement was given to the setting up of agricultural co-operatives, with the result that by 1972 there were 8,500 such co-operatives, covering 30,000 villages and with a membership of 8 million. One of the most important tasks of these co-operatives is, in concert with the Agricultural Credit Bank, to provide credit to farmers so as to enable them to make use of modern methods and machinery, and so as to free them from indebtedness to moneylenders.

The third phase of Land Reform, which is now beginning, aims at the adoption of modern techniques and the planned production of crops in accordance with national needs. Specifically it sets out to increase farm production so as to meet all the country's needs of food and raw materials; to raise the incomes and the standard of living of those who produce this food; and, by the adoption of the most efficient methods, to hold down prices to consumers. The achievement of those aims is to be helped by the creation of large farming units which will be formed by the voluntary amalgamation by individual farmers of their land, so as to form units of a size best adapted to modern farming.

By 1971 twenty such corporations had been set up, involving 106 villages with a population of over 53,000 people. In only a couple of years it was said that yields on these farms had risen by over 50 per cent.

While these changes were taking place on the purely farming side progress was being made in the fight against illiteracy. Before 1963 only 600,000 children in the rural areas attended school and only 4 per cent of all villages had any school at all. A Literacy Corps was formed, consisting of young graduates, who went out into the rural areas to teach. By 1971 there were members of this Corps teaching in over 20,000 villages, with 1½ million children and quarter of a million adults attending their classes. By 1972 more than 15,000 new village schools had been built.

A similar attack was made on rural health. In 1962 there were no more than 4,500 doctors in the whole of Iran, and of these only 135 were in villages. Of the 15 million people living in rural areas, only one million had access to any medical service. By 1970 this figure had risen to 8 million. The Health Corps was set up in 1964, consisting of doctors, dentists, veterinarians, and pharmacologists, drafted for military service but instead sent as mobile teams to serve in rural areas. Eight years later there were 400 such medical groups, each covering 20 to 40 villages. In that time infant mortality had decreased by 49 per cent and total mortality by 24 per cent. Rural areas have also received large sums for road building and various social services. Under the first and second development plans agriculture received 25 per cent of the total national investment, only slightly less than social affairs and communications, both of which give great benefits to rural areas.

Already the results of all these different measures can be seen. In 1965–66 the weight of field fodder produced was 500,000 tons: in 1972–73 it was 1,500,000 tons. In 1960–61 the wheat crop was 2·9 million tons: ten years later it was 3·8 million tons. The rice harvest had risen in the same period from 470,000 tons to 850,000 tons, cotton from 328,000 tons to 455,000 tons, and sugar beet from 700,000 tons to 3·9 million tons. Tractor sales had trebled in three years, and fertiliser consumption, which before 1963 was negligible, reached over 350,000 tons by 1973.

All this has been achieved in ten years. That is a long time for those who are hungry to wait for more to eat, but it is a short time in the context of the farming cycle. At the very least the Iranian experience shows the scale of progress that could be made in the world between the present time and the end of the century—if the will and the financial resources were there.

So far as Israel is concerned, even before the formation of the State of Israel in 1948 there was a tradition of co-operation among farmers in Palestine. This was fostered by the government of the new state, and was especially suited to the settlement on the land of newcomers from other countries. Because of the limited amount of land available, and the low rainfall, it was essential that the most efficient use should be made both of land and of water resources. It was also essential that efficient marketing should ensure that such food as was produced should reach the consumer, whether in Israel itself or in the countries to which the product was exported, with the minimum of waste. For exports especially, the need for high and consistent quality was paramount. To help bring this about the early governments undertook intensive research and provided widespread advisory services. They also instituted farm planning and provided investment loans and operating capital to enable the rapid adoption of modern techniques. At the same time Marketing Boards were set up, both for the home market and for export.

The government's attitude towards farming was based upon its belief that agriculture was one of the most important sectors in Israel's economy: it also held that agriculture had a vital part to play in building a new society. This was in marked contrast to most countries of the developed world, who looked to industry and the cities for the improvement of social and economic conditions, with agriculture and the countryside playing a very secondary role. The achievement of these objectives was greatly helped by the co-operative structure of much of Israel's agriculture. The most widely known form of co-operation was the *kibbutz*, the rural settlement which had existed long before the foundation of the state of Israel.

The *kibbutz* is in no way a co-operative farm. It is rather a settlement of people working together in a more or less tribal organisation, with shared responsibilities for the welfare of the entire group. The modern *kibbutz* cultivates the land of the settlement and produces either its own food or crops for export, or both: but it also has its own industries and its own professional and welfare services. Its basic tenets are equality, democracy, voluntary effort, and the satisfaction of the needs of its members by community rather than individual effort.

By 1973 there was some form of industrial activity in two out of every three *kibbutzim*, and the gross income of the *kibutzim* from this source had risen from an earlier figure of 25 per cent to 40 per cent. This rural industrialisation had become essential because of the rapidly rising productivity in agriculture, since there was less need for labour on the land: had industry not developed in the *kibbutz* many of the younger members would have been forced to seek their livelihood elsewhere. As it was, they now had the opportunity to remain in their own community, with the choice of a wider variety of jobs to suit their own particular abilities and preferences. The growth of industry in the *kibbutz* also ensured that there would be no great disparity of income or amenities between those working on the land, in factories, in commerce or the professions.

The *moshav* exemplifies a different type of co-operative settlement. It is based on independent work, and mutual aid. The *moshav* is granted a lease of its land from the government for 49 years, and in its turn grants a similar lease to the individual settler, whose right to the land is dependent upon his cultivating it. The size of the individual holding is the same for all members of the *moshav*, and depends upon the type of farming practised in each *moshav*. Every member cultivates the land himself, without outside help, and makes his own living out of his own work; but the community bears the responsibility for ensuring to each of its members a minimum subsistence, including social security and education for the children—always providing work or other obligations are not neglected.

The average *moshav* consists of 70 to 80 farms; together they account for 19,000 farms out of a total in the whole country of 35,000, with 100,000 people living on them, out of a total agricultural population of 307,000 (1971). As the crops grown on the *moshavim* become increasingly directed to export or consumption within Israel but away from the *moshav* itself, so have marketing facilities become more imporant. Neighbouring *moshavim* have therefore grouped themselves into co-operative associations for such things as slaughter-houses, cold storage, transport, and the purchase of supplies. As a result, members have the advantages of large-scale buying and selling, and the management committee of the *moshav* organises for its members such things as bank loans, irrigation water, and other services, as well as the supply of seeds and fertilisers and general farm requisites.

This original government policy and expenditure, operating largely through the *kibbutzim* and *moshavim*, but also through private farmers, large and small, has led to significant advances in the country's agricultural production during the first twenty-five years of the country's existence, from 1948 to 1973. The cultivated area has increased from 165,000 hectares to 422,500 hectares, and the irrigated area from 30,000 hectares to 179,000 hectares. Grain production has increased more than sixfold, from 55,000 tons to 360,000 tons; vegetables from 93,000 tons to 514,000 tons; citrus from 275,000 tons to 1,550,000 tons; meat from 7,500 tons to 195,000 tons; milk from 79,000 million litres to 467,000 million litres. In 1949 there were 680 agricultural tractors: in 1972 there were 14,500. In twenty years the consumption of fertilisers trebled, from 21,300 tons of nutrient to 60,000 tons.

In 1972 the export of farm produce earned $230 millions, which balances the amount spent on imported food and raw materials for agriculture: in other words Israeli agriculture had produced enough food and foreign exchange for its own nutritional needs. Israel has shown that, even with limited land and water, food production can be enormously increased, given the determination of the government to achieve success and to make available the necessary resources.

All those countries, in their different ways and with their differing political systems have demonstrated what can be done to stimulate agricultural production if the necessary pre-conditions are met. In all cases there has been a conscious and deliberate decision by the government to invest more money in agriculture on a long-term basis: this has involved expenditure on the infrastructure of the countryside in the shape of roads, communications, and, in some cases, irrigation.

It has also involved expenditure on the infrastructure necessary for improving the quality of life of the farmer, in the shape of schools, hospitals, housing, and services. It has involved investment in the factories needed to produce the equipment essential for modern farming, and in research, without which future progress is impossible. It matters little if this investment has come, as in the case of Iran, from indigenous resources derived from oil, or, as in the case of Cuba, from loans from other countries. The essential fact is that it is in agriculture and the rural infrastructure that the investment has been made.

In some cases this decision has involved a fundamental change in the pattern of land ownership and in all four countries the pattern of rural life has been drastically altered. This investment and these changes are sufficient evidence to convince the people of each of the countries that farming is an activity of value to their country, and that by engaging in it they will not be regarded in any way as inferior to those who embark on some other occupation, that they will receive material rewards comparable to those of urban workers, and that their families will have the same amenities and opportunities as they would have if they were brought up in the city.

The degree of success that these very different countries will eventually achieve cannot be assessed for many years to come; but at least there is evidence that the general proposals put forward earlier in this book are not merely theoretical, but in some form or another are already being implemented by countries of different history, different social composition, and different political complexions.

XIV

Summing Up

It is now time to put together the facts, the conclusions, and the proposals of the preceding chapters. First there is the fact that at the present time there are people in the world who are dying of starvation, and many more who are suffering to a greater or lesser extent from the effects of malnutrition. This is brought about in part by unequal distribution of the food that is being grown, in part by poverty, but above all because not enough food is being produced to meet the needs of all.

Second is the fact that this is no new state of affairs, but has been present in the world far back to the beginnings of history and beyond.

Third, as world population rises the situation will get worse if we continue on our present course and with our present sets of values.

From these facts several conclusions are drawn. On moral grounds alone it is wrong for some people in the world to have enough, and in certain cases more than enough, to eat when others either in the same country or elsewhere have too little. On economic grounds it is inefficient that many people who should be capable of producing food or other things that they and others want are unable to do so because of physical weakness, lack of education, or absence of the right tools. Because of malnutrition less total wealth is produced in the world. The standard of living not only of food producers but of those who produce manufactured goods is lower than it otherwise would be, because primary producers do not have the money to buy the products of the factories. On political grounds it is dangerous to divide the world into rich and poor, the haves and have-nots, the well-fed and the hungry. In the past such divisions have helped to bring about bloody revolution: in the future there is no reason for thinking that the same will not happen, but on a continental or world scale rather than a national one.

It is also concluded that, with the right incentives, more food could be produced from land already under cultivation. These incentives include a greater reward to those who cultivate the land, relative to the rewards obtained by people engaged in other activities; a diversion of resources from certain existing uses to agriculture, including research and the manufacture of inputs essential to modern farming; and an improvement in the quality of life in the country as compared with that of the cities. But the increased production achieved by such means would not be enough to produce all the food that is needed today, and will be needed in rapidly increasing quantities between now and the end of the century. The only hope of reaching the target lies in bringing into cultivation a great deal more land than is at present being farmed.

To achieve these aims various proposals are made. In the first place there must be a far greater investment in food production than there has been in the past. This investment will be partly in research and in the education of the farmer in the application of the results of that research: it will also be in the design and manufacture of machinery and in the production of fertilisers and herbicides. But this will not be enough. There must be massive investment in the reclamation of land that at the moment is producing nothing, in the drainage and irrigation of this land, in the construction of roads and other means of communication, in the building of schools and hospitals and the provision of electricity.

While this is being done a determined effort must be made to close the gap between town and country, between industry and agriculture. No longer must the city be regarded as the Mecca of all ambitious young people, a job in a factory or an office the ultimate symbol of success. Farms cannot be brought to the town; but factories can be brought to the countryside. A determined effort must be made in all developing countries to build new factories, whenever economic and technical factors allow, in country districts. In this way, as in the *kibbutzim* of Israel, those who are not needed on the land, or who do not have the qualities needed to become good farmers, can find well paid and useful employment while still living in or near the villages where they were born. A country where industry is spread throughout the countryside will be healthier, both economically and socially, than one where factories are concentrated in the big cities and the rural areas are devoted solely to agriculture.

Some of the money for this development will come from the consumers, who will in the future have to spend a larger proportion of their income on food than they have done in the past: but, while this should provide most of the money needed to redistribute wealth between those who produce food and the rest of the community, it will not provide the funds that are needed for investment. These can only come from the national budgets of the rich countries. A relatively small proportion of the amounts being spent today on defence and on space research will provide enough money for these purposes.

This money would be distributed in a variety of ways. Much of it should be made available to some international body such as the World Bank, which would be charged with the task of handing it over for specific projects. The newly constituted World Food Authority would play a key part in this. It would draw up plans, solely on grounds of efficient use of resources, for improving the productivity of land already under cultivation and for bringing fresh land into cultivation. The plans would include the means of improving the amenities of life in the rural areas so as to encourage the most intelligent and go-ahead people to remain on the land, or even to leave the cities and take up farming.

All these proposals entail a transfer of real wealth from rich countries to poor, and, to a lesser extent, from rich, or relatively rich, people to poor people. Here, unpopular though it may be, it must be made clear that the worker in the USA or in Western Europe, earning his $200, his 300 marks, or his £40 per week, although he considers himself poor in relation to those he sees around him in his own country, is rich compared with the millions in the developing countries who have an *average* income of less that $200 a year. Where this is the average there are very many whose incomes are even lower. According to figures published by the World Bank there were in 1970 1,000 million people with an average annual income of $105, and it is estimated that by 1980 this will have risen to no more than $108. In the OECD countries, on the other hand, the 1970 figure was $3,000, with an estimated increase to $4,000 by 1980.

Some of this transfer of wealth would be by way of investments which, in the long run, would yield commercial returns. At the outset, however, such investment is rarely attractive to existing enterprises. The amount of money involved is large, and many years must elapse before profits are earned. Good management is hard to come by, and there may be political as well as purely business risks. To stimulate such investment and to minimise these handicaps, legislation was enacted in the UK which resulted in the setting up of what was then called the Colonial Development Corporation, subsequently changed to Commonwealth Development Corporation, now with powers to operate in developing countries outside the Commonwealth. The Corporation's activities fill the gap between government aid, which is mainly given for infrastructure, and purely private enterprise operations.

The Corporation's funds came originally, and still come to a large extent, from the British government, but interest has to be paid on these sums, though at a rate lower than commercial bank borrowings; and the capital has eventually to be repaid. When the project—be it a rubber plantation, a pulp mill, an electricity supply undertaking or a housing development—starts to earn money it may be taken over, in whole or in part, by private investors or the government of the country in which it is situated, and the capital received for it will, after repayment of the British government loan, be available for new investment.

A massive extension of similar types of organisation, on national or regional bases, will be needed in the future to ensure that enterprises, which eventually will become commercially profitable, are set up sufficiently quickly.

There will, however, always be the need for heavy investment in projects which will never yield a direct commercial return. Schools, hospitals and roads will never show a return on capital which would satisfy a private investor. As an example of this it was calculated some fifteen years ago that the cost of building, at then ruling prices, all existing schools and other places of learning, hospitals, roads and railways in the UK amounted to some £15,000 million. If interest alone, at five per cent, had to be paid on this capital sum it would be equivalent to a charge of £15 per year on every man, woman and child in the UK. At today's costs and interest rates this figure would be increased by between five and ten times.

Even accepting the fact that a poor developing country would not need services approaching the standard of the UK today, the burden of interest charges on a population with an average annual income per head of £50 would manifestly be insupportable. If such services are to be made available, as they must, to these countries, it can only be by way of gifts.

The transfer of wealth, therefore, must be brought about in three ways. First, by an increase in the price that the consumer pays for his food, relative to the rest of his expenditure. This will encourage more food production in all countries, developed and under-developed. Second, by making available, on an international basis, investment capital for pump-priming operations, this investment eventually yielding a commercial return. Third, by direct gifts from the rich to the poor countries of the money needed for the infrastructure without which their economic growth, and in particular their food production, cannot increase at the pace which the world needs.

The World Food Authority, working with the Food and Agriculture Organisation, would also draw up a World Food Plan, and seek the co-operation both of producer and consumer countries in operating it. It would be based on long-term needs and long-term production potential, and would be implemented by long-term contracts between producers and consumers, either on a government-to-government basis, or through Marketing Boards or similar bodies. An essential part of this plan would be that sufficient money should be available to enable food surpluses to be bought as they arise, so that prices do not collapse, with consequent loss of confidence and investment power of the producer. In the case of non-perishable products these surpluses would be stored and released at times of shortage, so as to prevent an undue rise in price; or made available to countries or areas where there is risk of starvation, at prices subsidised out of the general fund of the Authority. In the case of perishables, these would be distributed quickly at subsidised prices where the need is greatest.

Proposals are also made for more orderly marketing and processing at the producer level, for the provision of credit with which to finance the investment needed for efficient production, and for the encouragement of co-operative effort for these purposes and for helping the rapid adoption of improved techniques. The importance of choosing the right system of landownership is also stressed, and especially the need of the farmer for security of tenure and for the provision of the right sort of permanent equipment, such as buildings and drainage.

All these proposals must of necessity be in very general terms. Different countries, and different areas of the same country, have different needs, based not only on soil and climate but on tradition, history and personal factors.

Many things have been omitted from the book. But it is hoped that the general outline of policy that it contains will at least be sufficient to stimulate thought among those who have the responsibility of preparing plans for their own country, as well as among those who, while consumers and not producers, are concerned that there should still be millions of starving people living in the world at the end of the 20th century.

One thing at least cannot be gainsaid. Unless rapid and drastic action is taken now to produce more food and to ensure its better distribution, those millions will, by the beginning of the 21st century, increase tenfold.